AN EARTHED FAITH:
Telling the Story amid the "Anthropocene"
Volume 3

THE PLACE OF STORY AND THE STORY OF PLACE

The Place of Story and the Story of Place
An Earthed Faith: Telling the Story amid the Anthropocene, volume 3

Originally published by AOSIS Books, an imprint of AOSIS Publishing

Copyright © 2025 Ernst M. Conradie. All rights reserved. Except for brief quotations in critical publications or reviews, no part of this book may be reproduced in any manner without prior written permission from the publisher. Write: Permissions, Wipf and Stock Publishers, 199 W. 8th Ave., Suite 3, Eugene, OR 97401.

Pickwick Publications
An Imprint of Wipf and Stock Publishers
199 W. 8th Ave., Suite 3
Eugene, OR 97401

www.wipfandstock.com

PAPERBACK ISBN: 979-8-3852-4317-4
HARDCOVER ISBN: 979-8-3852-4318-1
EBOOK ISBN: 979-8-3852-4319-8

Cataloguing-in-Publication data:

Names: Conradie, Ernst M. | Pan-chiu, Lai

Title: The place of story and the story of place: An earthed faith: telling the story amid the anthropocene, volume 3 / Ernst M. Conradie and Willie James Jennings.

Description: Eugene, OR: Pickwick Publications, 2024. | AThe place of story and the story of place: Telling the Story amid the Anthropocene. | Includes bibliographical references and index.

Identifiers: ISBN 979-8-3852-4317-4 (paperback). | ISBN 979-8-3852-4318-1 (hardcover). | ISBN 979-8-3852-4319-8 (ebook).

Subjects: LCSH: Theology. | Human ecology. | Ecotheology. | Trinity. | Storytelling.

Classification: BR115 T10 2025 (print). | BR115 (ebook).

AN EARTHED FAITH:
Telling the Story amid the "Anthropocene"
Volume 3

The Place of Story and the Story of Place

Editors
Ernst M. Conradie
Willie James Jennings

PICKWICK Publications · Eugene, Oregon

Theological and Religious Studies editorial board at AOSIS

Chief Commissioning Editor: Scholarly Books
Andries G van Aarde, MA, DD, PhD, D Litt, South Africa

Board Members
Chen Yuehua, Professor of the School of Philosophy, Zhejiang University, Hangzhou, China.
Christian Danz, Professor of the Institute for Systematic Theology and Religious Studies, Evangelical Theological Faculty, University of Vienna, Vienna, Austria.
Corneliu C Simut, Professor of Historical and Systematic Theology, Faculty of Theology, Department of Theology, Music and Social-Humanistic Sciences, Emanuel University of Oradea, Romania; Supervisor of doctorates in Theology, Interdisciplinary Doctoral School, Aurel Vlaicu State University of Arad, Romania; Associate Research Fellow in Dogmatic Theology, Faculty of Theology, Department of Dogmatics and Christian Ethics, University of Pretoria, South Africa.
David D Grafton, Professor of Islamic Studies and Christian-Muslim Relations, Duncan Black Macdonald Center for the Study of Islam and Christian-Muslim Relations, Hartford International University for Religion and Peace, United States of America.
David Sim, Professor of New Testament Studies, Department Biblical and Early Christian Studies, Australian Catholic University, Australia.
Evangelia G Dafni, Professor of School of Pastoral and Social Theology, Department of Pastoral and Social Theology and Department of the Bible and Patristic Literature, Faculty of Theology, Aristotle University of Thessaloniki, Greece.
Fundiswa A Kobo, Professor of Department of Christian Spirituality, Church History and Missiology, University of South Africa, South Africa.
Jean-Claude Loba-Mkole, Professor of Hebrew and Bible Translation, Department of Hebrew, Faculty of Humanities, University of the Free State, South Africa.
Jeanne Hoeft, Dean of Students and Associate Professor of Pastoral Theology and Pastoral Care, Saint Paul School of Theology, United States of America.
Lisanne D'Andrea-Winslow, Professor of Department of Biology and Biochemistry and Department of Biblical and Theological Studies, University of Northwestern-St. Paul, Minnesota, United States of America.
Llewellyn Howes, Professor of Department of Greek and Latin Studies, University of Johannesburg, South Africa.
Marcel Sarot, Emeritus Professor of Fundamental Theology, Tilburg School of Catholic Theology: Religion and Practice, Tilburg University, the Netherlands.
Nancy Howell, Professor of Department of Philosophy of Religion, Faculty of Theology and Religion, Saint Paul School of Theology, Kansas City, United States of America.
Piotr Roszak, Professor of Department of Christian Philosophy, Faculty of Theology, Nicolaus Copernicus University, Poland.
Sigríður Guðmarsdóttir, Professor of Department of Theology and Religion, School of Humanities, University of Iceland, Reykjavík, Iceland; Centre for Mission and Global Studies, Faculty of Theology, Diakonia and Leadership Studies, VID Specialized University, Norway.
Wang Xiaochao, Dean of the Institute of Christianity and Cross-Cultural Studies, Zhejiang University, Hangzhou, China.
Warren Carter, LaDonna Kramer Meinders Professor of New Testament, Phillips Theological Seminary, Oklahoma, United States of America.
William RG Loader, Emeritus Professor of New Testament, Murdoch University, Western Australia.

Peer-review Declaration

The publisher (AOSIS) endorses the South African "National Scholarly Book Publishers Forum Best Practice for Peer-Review of Scholarly Books." The book proposal form was evaluated by our Theological and Religious Studies editorial board. The manuscript underwent an evaluation to compare the level of originality with other published works and was subjected to rigorous two-step peer-review before publication by two technical expert reviewers who did not include the volume editor and were independent of the volume editor, with the identities of the reviewers not revealed to the editor(s) or author(s). The reviewers were independent of the publisher, editor(s) and author(s). The publisher shared feedback on the similarity report and the reviewers' inputs with the manuscript's editor(s) or author(s) to improve the manuscript. Where the reviewers recommended revision and improvements, the editor(s) or author(s) responded adequately to such recommendations. The reviewers commented positively on the scholarly merits of the manuscript and recommended that the book be published.

Research Justification

Christian ecotheology is at times reduced to creation theology, anthropology or environmental ethics. Nevertheless, adequate creation theologies that do justice to the materiality of what is created (*creatura*), to the act of creating (*creatio*) and the identity and character of the Creator, remain surprisingly scarce. Moreover, the danger that creation theologies tend to legitimize positions of power and privilege remains prevalent amid the legacy of imperialism, colonialism and apartheid. In the context of the "Anthropocene," where a clear separation between nature and culture can no longer be condoned, an adequate creation theology has become urgent. This third volume of the "An Earthed Faith: Telling the Story amid the 'Anthropocene'" series captures the state of the debate in contemporary ecotheology on creation theology and extends this debate through a set of diverse contributions from around the world.

The title of the volume suggests that creation narratives necessarily emerge from within a particular context that requires a sense of place—one that can come to terms with the destructive dynamics of power in that place. This indicates a tension between place and displacement and allows for a dialectic of orientation, disorientation and reorientation. At the same time, there is a need to recognize cosmic, biological and cultural evolution by telling the story of any particular place. The question, then, is how such a story is to be told amid the rupture associated with the "Anthropocene."

The ambitious aim of the editorial introduction to this volume is to describe the state of the debate on creation theology, especially (but not only) in Christian ecotheology. It does so by outlining how the debate has shifted from one agenda to another over the last five decades. It concludes that there remains considerable confusion on what the question is that creation theology addresses: Should the focus be on whether, how, who, what, why or for what purpose the world was created, or what the act of creating may mean? On this basis, each of the essays then addresses the following core question raised in this volume: "What difference does it make to the story of cosmic, planetary, human and cultural evolution to re-describe this as the creative work of God's love? Inversely, what difference does it make to the story of God's love to describe it in evolutionary and geographic terms?"

The essays included in this volume are all original and develop constructive responses to the same underlying question from within distinct contexts. They adopt a similar methodology, namely, a critical and constructive review of the available literature in the field of Christian ecotheology. These are scholarly essays in the sense that they are written by leading scholars together with a few emerging scholars in the field. The volume is aimed primarily at experts in the field and has been checked for plagiarism and self-plagiarism.

The ten contributors (two had to withdraw belatedly) were selected in order to optimize a diversity of positions in terms of geographical context, confessional traditions, and theological schools, also taking considerations of gender, race, age and language into account.

The concluding conversation between the contributing authors identifies current paths and emerging horizons in creation theology in order to take the debate forward in the context of contemporary ecotheology. It invites other participants in the field to join the conversation on the basis of this volume.

Ernst M. Conradie, Department of Religion and Theology, University of the Western Cape, Bellville, South Africa.

Willie James Jennings, Yale Divinity School, Yale University, New Haven, Connecticut, United States of America.

Artist statement

Garth Erasmus is a South African artist born in Uitenhage and currently based in Cape Town. He participated, together with Nesindano Xhoes Namise, Peter Thiessen and Ruth May, in an exhibition entitled "Sand Sand Sand" at the Kunsthaus Hamburg in Hamburg, Germany, from 20 August to 02 October 2022. A unique feature of this exhibition is that the artworks were not displayed on the wall but on the floor as sand work or hanging from the roof. Ink on Paper 2018 Drawing 7 was one such work.

Erasmus explains the significance of this exhibition in the following way:

> 'In the frame of an international collaboration, visual artists and musicians explore sand – as a material and as a medium for reflecting on traces of colonial history, for the reappropriation of indigenous knowledge and the respective worldviews. Sand is like the skeleton of the world, a foundation of modern culture and technologies, the invisible but essential ingredient of constructed living environments and digital communication devices. For its extraction, entire beaches are hauled away, seabeds are sucked out, gigantic pits are dug out, mountains are piled up. Wars are waged, villages sink into the ground and islands are washed over by the sea. Against this background, the artists participating in the project will create a space at the Kunsthaus Hamburg, in which fleeting sounds, fabric, combined with textile and mineral images interweave in a spatial and acoustic installation to form a landscape.'

For more detail on this exhibition see https://www.mutualart.com/Exhibition/Sand-Sand-Sand/75D00D60799FB126.

Table of Contents

Abbreviations and Acronyms Appearing in the Text and Notes	xiii
Notes on Contributors	xv
An Earthed Faith: Envisaged Volumes in the Series	xix

The Place of Creation in Christian Ecotheology – Some Shifts in the Story — 1
Ernst M. Conradie with Willie James Jennings

The Series on "An Earthed Faith"	1
The Question Posed in this Volume	3
Agenda 1: The Self-critique of (Western) Male Theologies	4
Agenda 2: Widening the Scope of Ecotheology Beyond Creation Theology	6
Agenda 3: Creation, the Natural Sciences, and Ecological Concerns	8
Agenda 4: The Shift to (Western) Ecofeminist Theologies of Creation	11
Agenda 5: Liberation, Black, Womanist, and Subaltern Creation Theologies	12
Agenda 6: The Retrieval of Indigenous Theologies of Creation Globally	16
Agenda 7: Ecumenical Efforts Toward Telling the Story Together	17
Conclusion: The Question is … What is the Question?	20
Bibliography	21

Treaty as a Shared Narrative: Indigenous Treaty as Canada's Creation Story — 25
Ray Aldred

Introduction	25
Heal the Land	27
Develop or Raise Emotional Intelligence and Skill	28
Cast a Vision of Unity That is Neither Tribalism nor a New Imperialism	29
Identity, Spirituality, and Law from Creation Stories	30
Identity	30
Spirituality	33
Law	34

Table of Contents

A Treaty Is an Attempt by First Nations at a Shared Narrative	35
Spirituality is Built upon the Understanding of a Good World	37
Conclusion	39
Bibliography	39

From an Island of Africa to an Island of Europe: Perspectives on a Theology of Creation from Madagascar and Crete — 41
Louk Andrianos Andriantiatsaholiniaina

Introduction	41
Sacredness and Virtue in God's Creation in the Malagasy Context	43
The Origin of Human Virtue in Malagasy Belief	44
Stories of *Fihavanana*—the Malagasy *Ubuntu*—and the Creation of the Island	45
A Story of the Youth "Blue Cross" Movement to Protect Creation	47
A Story of the "Mirror" at the Orthodox Academy of Crete to Praise the Creator	49
Feminine Influence in Understanding an Orthodox Theology of Creation	51
Living a Theology of Creation with the Ecumenical Patriarchate of Constantinople and the World Council of Churches	53
The Greed Line Study	55
Conclusions	57
Bibliography	58

Ecotheology and Creation Theology: Shona People's Indigenous Cosmologies — 59
Sophia Chirongoma

Introduction and Self-Location	59
The Mwedzi/Dzivaguru Creation Myth	61
The Guruuswa Creation Myth	65
Interrogating Inherent Gender and Social Inequalities Within Shona Society	67
The Encroaching Ecological Crisis in Zimbabwe	68
The Interface between Creation Theology and a Theology of Responsible Ecological Stewardship	71
Conclusion	75
Bibliography	76

Reforming Place or Placing Reform? One Western Cape Version of the Story   79
Ernst M. Conradie

Situating Creation Theology	79
The Places of My Story	81
The Reformed Stories of Place That Shaped Me	83
Storying Place	87
Placing Story: A Constructive Thesis	91
Seeing the World as God's Creation: The Difference That Makes a Difference	96
Bibliography	98

The Story of a Place in the North: Natural Disasters within God's Good Creation—A Lutheran Perspective   101
Arnfríður Guðmundsdóttir

My Place of Origin	103
A Pastor Serving God and Distressed People in Catastrophic Circumstances	105
"The Destructive Fires of God"	107
To be Created and Cared for by a Loving God	110
Prayers of Lament	111
When the Creative Work of God's Love Is Destroyed by Natural Disasters	112
Eruption in Heimaey	113
Avalanches in the Westfjords	114
"Christs to One Another"	116
Conclusion	117
Bibliography	118

Place and Space: Gathering Wisdom from the Life Work of Proto-Ecowomanist Fannie Lou Hamer   121
Melanie L. Harris

Introduction	121
Origins	122
African Cosmology and African American Stories of Place and Space	124
Fannie Lou Hamer: Place and Space	127
Bound with the Earth: Fannie Lou Hamer Biographic Notes and Hamer's Courageous Contributions to Climate Justice	128

Hamer's Theo-Ethic of Love: Three Movements	131
Conclusion	135
Bibliography	135

An Arawete Cannibal Theology of Noncreation: Patriarchy, Capitalism, and Food   137
Eneida Jacobsen

Placing My Essay	137
An Arawete Theology of Creation in Between Places	140
Patriarchal Societies and Patriarchal Theologies	145
An Arawete Critique of Capitalism: From Patriarchy Back to Nature	148
Conclusion	151
Bibliography	153

The Struggle of Earth-Storytellers   155
Willie James Jennings

Garden Life	155
Displacement	158
The Body-Land Connection	161
Bibliography	168

Continuing the Conversation on Creation in Christian Ecotheology   171

Index   185

Abbreviations and Acronyms Appearing in the Text and Notes

COP	Conference of Parties
ECOTHEE	Establish a Biannual Conference on Ecological, Theological, and Environmental Ethics
FJKM	*Fiangonan' i Jesoa Kristy eto Madagasikara*
FORTH	Foundation for Research and Technology, Hellas
IKS	Indigenous Knowledge Systems
ITHE	Institute for Theology and Ecology
OAC	Orthodox Academy of Crete
SIDS	Small Island Developing States
SNCC	Student Nonviolent Coordinating Committee
SOC	Season of Creation
TRCA	Truth and Reconciliation Commission of Canada
UWC	University of the Western Cape
WCC	World Council of Churches

Notes on Contributors

Ray Aldred is status Cree from Swan River Band, Treaty 8, in Northern Alberta, Canada. Northern Alberta forms part of Treaty 8, the historic numbered treaties signed in Canada. Born in Northern Alberta, he now resides with his wife in Richmond, British Columbia. Reverend Ray Aldred is a husband, father, and grandfather. He is the director of the Indigenous Studies Program at the Vancouver School of Theology, whose mission is to partner with the Indigenous church around theological education. He was first ordained with the Christian and Missionary Alliance in Canada and is now ordained with the Anglican Church of Canada. His PhD thesis from the University of Toronto is entitled "An Alternative Starting Place for an Indigenous Theology" (2020).

He is registered as a coresearcher at the University of the Western Cape for the project on "An Earthed Faith: Telling the Story amid the 'Anthropocene.'"

ORCID: https://orcid.org/0009-0003-5336-5476
Email: raldred@vst.edu

Louk A Andrianos is a World Council of Churches consultant on the Care for Creation, Sustainability, and Climate Justice. He was born in Madagascar and lives permanently on Crete, Greece. After his Master's Degree in Hydrology (Brussels), he received a PhD in Sustainable Development Sciences from the Technical University of Crete and a second Master's Degree in Plant Molecular Biology. He is head of the Institute of Theology and Ecology at the Orthodox Academy of Crete. He is the author of *The Fuzzy Limiting Factors for Sustainable Development* (2008), *The "Ecumenical Theology of Hope for the Common Oikos* (2019), and co-editor of *Kairos for Creation: The Wuppertal Call* (2019), *Contemporary Ecotheology, Climate Justice and Environmental Stewardship in World Religions* (2021), and *Penser les relations écologiques en théologie à l'ère de l'anthropocène"* (2023).

He is a member of the Royal Institute Advisory Council and is registered as a coresearcher at the University of the Western Cape for the project on "An Earthed Faith: Telling the Story amid the 'Anthropocene.'"

ORCID: https://orcid.org/0009-0003-4796-0594
Email: louk.andrianos@wcc-coe.org, luc_andrian@yahoo.gr

Sophia Chirongoma is a senior lecturer in the Religious Studies Department at Midlands State University, Zimbabwe. She is also an Academic Associate/Research Fellow at the Research Institute for Theology and Religion (RITR) in the College of Human Sciences, University of South Africa. Her research interests and publications focus on the interface between culture, ecology, religion, health, and gender justice.

ORCID: https://orcid.org/0000-0002-8655-7365
Email: sochirongoma@gmail.com or sochirongoma@yahoo.com;

Ernst M Conradie is a senior professor in the Department of Religion and Theology at the University of the Western Cape in South Africa. He works in the intersection between Christian ecotheology, systematic theology, and ecumenical theology and comes from the Reformed tradition. He is the author of *The Earth in God's Economy: Creation, Salvation and Consummation in Ecological Perspective* (2015), *Redeeming Sin? Social Diagnostics amid Ecological Destruction* (2017), and *Secular Discourse on Sin in the Anthropocene: What's Wrong with the World?* (2020). He was the international convener of the Christian Faith and the Earth project (2007–2014), and co-editor with Hilda Koster of *The T&T Clark Handbook on Christian Theology and Climate Change* (2019). He is responsible for registering the project on "An Earthed Faith: Telling the Story amid the 'Anthropocene'" at the University of the Western Cape.

ORCID: https://orcid.org/0000-0002-0020-6952
Email: econradie@uwc.ac.za

Arnfríður Guðmundsdóttir is a professor of Systematic Theology, Faculty of Theology and Religious Studies, University of Iceland. She was educated at the University of Iceland, University of Iowa, University of Chicago, and Lutheran School of Theology at Chicago. She is the author of *Meeting God on the Cross: Christ, the Cross, and the Feminist Critique* (Oxford University Press, 2010) and has written articles and book chapters in Icelandic and English within the fields of Christology, Lutheran theology, feminist theology, ecofeminism and climate justice, as well as religion and film. She is the editor of *Studia Theologica Islandica* and *Studia Theologica: Nordic Journal of Theology* and a co-editor of the T&T Clark Explorations in Theology, Gender and Ecology book series. She is an ordained pastor within the Evangelical Lutheran Church of Iceland.

She is registered as a coresearcher at the University of the Western Cape for the project on "An Earthed Faith: Telling the Story amid the 'Anthropocene.'"

ORCID: https://orcid.org/0000-0001-6432-1434
Email: agudm@hi.is

Melanie L Harris is the director of Food, Health, and Ecological Well-Being and professor of Black Feminist and Womanist Theologies, jointly appointed with African American Studies and the School of Wake Forest Divinity at Wake Forest University. Formerly associate dean of Diversity, Equity, and Inclusion with AddRan College at Texas Christian University, her leadership, teaching, research, and scholarship focus on the areas of Religious Social Ethics, Environmental Justice, Womanist Ethics, and African American Religion. She is the author of *Ecowomanism: African American Women and Earth Honoring Faiths*, *Ecowomanism, Religion and Ecology*, *Gifts of Virtue: Alice Walker and Womanist Ethics* and co-editor of the volume *Faith, Feminism, and Scholarship: The Next Generation* with Kate M. Ott. She is currently writing two books engaging ecowomanist meditations and sermons, and the proto-ecowomanist activism of Harriet Tubman, Fannie Lou Hamer, and Alice Walker.

She is registered as a coresearcher at the University of the Western Cape for the project on "An Earthed Faith: Telling the Story amid the 'Anthropocene.'"

ORCID: https://orcid.org/0009-0003-5214-7784
Email: drmelharris@gmail.com

Eneida Jacobsen, or Aeneid Jacob, is an adjunct professor and PhD student in the Department of Philosophy at Villanova University, Philadelphia. Jacobsen holds a PhD in Theology from the Lutheran School of Theology in São Leopoldo, Brazil. They are co-editor of *Public Theology in Brazil: Cultural and Social Challenges* (LIT Verlag, 2013), and author of *Theologie und politische Theorie: Kritische Annäherungen zwischen zeitgenössischen theologischen Strömungen und dem politischen Denken von Jürgen Habermas* (Peter Lang, 2018). Their research interests include public theology, liberation theology, social and political philosophy, Marxism, feminism, and Amerindian perspectivism. Jacobsen is registered as a coresearcher at the University of the Western Cape for the project on "An Earthed Faith: Telling the Story amid the 'Anthropocene.'"

ORCID: https://orcid.org/0009-0003-9171-135X
Email: eneida.jacobsen@villanova.edu, eneida.jacobsen@yahoo.com.br

Willie James Jennings is a professor of Theology and Africana Studies at Yale University Divinity School. Educated at Calvin College (now University), Fuller Theological Seminary, and Duke University (all in the United States), he was formerly the academic dean at Duke Divinity School. He is the author of several award-winning books, including *The Christian Imagination: Theology and the Origins of Race* (Yale Press, 2011), *A Commentary on the Book of Acts* (WJK Press, 2017), and *After Whiteness: An Education in Belonging* (Eerdmans, 2020). His research interests include creation, ecological studies, the built environment, critical geography, critical race theory, liberation theology, and decolonial thought. His current writing project is a doctrine of creation.

He is registered as a coresearcher at the University of the Western Cape for the project on "An Earthed Faith: Telling the Story amid the 'Anthropocene.'"

ORCID: https://orcid.org/0009-0008-9462-117X
Email: willie.jennings@yale.edu

An Earthed Faith: Envisaged Volumes in the Series

The following twelve volumes are envisaged in the series entitled "An Earthed Faith: Telling the Story amid the 'Anthropocene'":

■ 1) Taking a Deep Breath for the Story to Begin ... An Earthed Faith 1 (Prolegomena)

This volume will address the following question: How does the story of who the Triune God is and what this God does relate to the story of life on Earth? Is the Christian story part of the earth's story or is the earth's story part of God's story, from creation to consummation? This raises many issues on the relatedness of religion and theology, the place of theology in multidisciplinary collaboration, the notion of revelation, the possibility of knowledge of God, hermeneutics, the difference between natural theology, and a theology of nature, etc. The word "breath" in the title suggests the Spirit of God as the source of inspiration for the story, already present in any further deliberations. It hints at an air of anticipation, indicated by the three dots in the title.

■ 2) How Would We Know What God is up to? An Earthed Faith 2 (Method)

This volume will address the following question: given what we know about the "Anthropocene," how does one even begin to answer the question of what is this God up to? And how would we know how to respond to that? These are questions of theological method, including the sources and interlocutors of Christian theology, its aims and starting points, social theories shaping it, and presuppositions grounding it. Addressing these questions is the classic task of doing contextual theology, namely, to describe and analyze the particular context that is addressed and to consider how this may best be addressed theologically. This question highlights the need for prophetic theology to discern the "signs of the times," to recognize a "moment of truth" (*kairos*), and to discern counter-movements of the Spirit. Such methodological questions are necessary in order to tell the story of who God is and what God does amid the "Anthropocene." In terms of narrative/rhetorical theory, a focus on method requires attention to the plot upon which the narrative hinges; the sense of crisis that will draw together the characters; and the exigencies that invite passion, reflection, and persuasion. Theological method is inherently a theological question about sin and salvation, creation and redemption, God and God's world—and shapes where the story may lead and how it may be told.

■ 3) The Place of Story and the Story of Place? An Earthed Faith 3 (Creation)

This volume will address the following question: "What difference does it make to the story of cosmic, planetary, human and cultural evolution to re-describe this as the creative work of God's love?" Inversely, what difference does it make to the story of God's love to describe it in evolutionary terms? Addressing this question will require theological reflection on creation and cosmic, biological, hominid, and human evolution (the story of place). Such reflection on the beginning is, of course, not situated "in the beginning" but entails a narrative reconstruction of the story where current interests, positions of power, and fears are necessarily at stake (the place where the story is being told). This is a contested space, indeed a "site of struggle," often dominated by issues of race rather than by grace. How, then, is this story to be told given a sense of place? It will not be possible to avoid questions around suffering, sin, evil, and the tragic (the theme of the next volume), but the focus will be on why on Earth a loving God would deem this story to be "very good"—despite the prevalence of suffering, injustice, and oppression?

■ 4) Making Room for the Story to Continue? An Earthed Faith 4 (Providence)

This volume will address the following question: How could the suffering of God's creatures in the "Anthropocene" be reconciled with trust in God's loving care? Addressing this question will require theological reflection on the classic themes related to the doctrine of providence, including *creatio continua*, *conservatio*, *gubernatio* and *concursus*. For some, God's providence (common grace) is a necessary requirement to allow (to make room for) the history of salvation to proceed. For others, the suffering embedded in God's "good" creation requires responses to the theodicy problem: Why would a loving God allow creatures to suffer so much? What is the relationship between so-called natural evil and social evil? Is the underlying problem human sin, or is it the inadequacies, the tragic dimension, indeed the violence embedded in God's world? Again, this last question is hinted at in the question mark after the title.

■ 5) The Saving Grace of the Story? An Earthed Faith 5 (Soteriology)

This volume will address the following question: How is the Christian message of salvation to be interpreted given current ecological destruction and apocalyptic fears associated with the "Anthropocene"? Is this message plausible given the failure of Christianity to address so many other urgent problems over 20 centuries? This will require theological reflection on Christological symbols such as atonement and Pneumatological symbols such as liberation, healing, reconciliation, regeneration, moral guidance, justification, and sanctification—insofar as these may be pertinent in the Age of the "Anthropocene." The title is ambiguous and ironic to indicate that the story is highly contested but is, at best, to be understood as good news for the whole Earth.

■ 6) The Keepers of the Story? An Earthed Faith 6 (Ecclesiology)

This volume will address the following question: What is the place and significance of the church in God's "household," now situated in the destabilizing context of the "Anthropocene"? Addressing this question will require theological reflection on the formation, up-building and very nature of the church, and on its many ministries and missions. Presumably, the question is no longer whether there is salvation outside of the church but indeed whether there is salvation to be found within the church. Can it still be said that the church is God's main (even only) instrument (sign, sacrament, icon) to bring salvation, given the challenges posed by the "Anthropocene"? Or is the task of the church the monastic one of "keeping" the story, that is, to maintain the inner secret to the mystery of history, amid dark clouds looming and despite few outsiders taking any notice? Does this not sound as if it is the church that needs to come to God's rescue, or is the inverse true?

■ 7) Where the Story Ends and its Ends ... An Earthed Faith 7 (Eschatology)

This volume will address the following question: How should the content and significance of Christian hope be understood in the context of the "Anthropocene"? Addressing this question will require theological reflection on the eschatological symbols of the final judgment as a sign of hope, on the resurrection of the dead, on the coming reign of God and on eternal life. It will also have to assess whether such hope is to be understood as the restoration (neo-Calvinism), elevation (Roman Catholicism), replacement (Anabaptism), recycling (liberalism/secularism) or deification/theosis (Eastern Orthodoxy) of this world. Does the meaning of the story lie in its end, or in the journey/pilgrimage toward that end? Any answer to such questions will remain provisional, hinted at in the three dots in the title.

■ 8) Being Blessed as the Inner Logic of the Story? An Earthed Faith 8 (Election)

This volume will address the following question: Can the notion of being God's chosen people/instrument be retained in a religiously plural world under the threat of the "Anthropocene"? Addressing this question will require theological reflection on the themes of divine election and vocation. Can "being blessed" by God be understood as the inner logic of the story? Is such blessing not often experienced as a curse? What about divine reprobation, punishment, and justice for the victims and perpetrators of history? How is a theology of religions to be understood in a context characterized by common threats, the need for tolerance, and compassion across religious divides? How can Christians move beyond the options of exclusivism and relativism in the context of the "Anthropocene"? What does it mean to be blessed and for the whole of creation to receive God's blessing?

■ 9) The Spirit of the Story? An Earthed Faith 9 (Pneumatology)

This volume will address questions around the identity and character of God's Spirit. It will require theological reflection on how the very notion of spirit should be understood in relation to person, matter, ideas, force, energy, and related concepts. What does it mean that this Spirit is "holy" and makes things "holy"? Is this Spirit able to overcome what is "demonic" in the "Anthropocene"? Is it money or love that makes the world go round? Or is this Spirit the spirit that makes matter move, even if this movement is not all that obvious and requires discernment?

■ 10) The Letter of the Story? An Earthed Faith 10 (Christology)

This volume will address questions around the identity and character of Jesus of Nazareth, proclaimed to be the Christ, anointed by God's Spirit, the One who would inaugurate God's coming reign. It will require theological reflection on the significance of all six Christological symbols, namely (deep) Incarnation, Cross, Resurrection, Ascension, Session, and Parousia as these may relate to the coming of the "Anthropocene." If the cross is a concrete symbol of the history of imperialism and oppression, can the (bodily?) resurrection still function as an equally concrete symbol of hope in the "Anthropocene"? How is the interplay between the letter and the spirit of the story to be understood given long-standing ecumenical divides on the *filioque* controversy—which still divides the East and the West, the North and the South—over whether the Spirit works (only/primarily) on the basis of the Letter (as most so-called mainline churches assume)? Or should the relative independence of God's Spirit be emphasized (as many others emphasize)?

■ 11) In Communion with the Storyteller(s)? An Earthed Faith 11 (Trinity)

This volume will address questions around the doctrine of the Trinity as the inner secret/apophatic mystery/doxological culmination of the Christian faith. It will offer theological reflection on how the economic Trinity and the immanent Trinity are related by exploring God's identity and character. The question is which of God's characteristics need to be foregrounded in the Age of the "Anthropocene." In particular, how is God's mercy related to God's justice, given the interactions between God as Father, Son, and Spirit? Can these (patriarchal) symbols be maintained in the "Anthropocene"? Should one favor the social analogy (emphasizing communion) or the psychological analogy (perhaps allowing for a more generic notion of God) for understanding the Trinity? What difference does faith in such a God make (if any) in the Age of the "Anthropocene"? Moreover, who is telling the story? Are we (Christians?) the ones responsible for telling the story or are we characters in a story ultimately told by Godself? Given these reflections, what does it mean to believe in "God" (a God, any God) in the world in which we now live? Note that this (philosophical) question is not addressed upfront but penultimately. For Christians, the question remains whether this Triune God can be regarded as the ultimate mystery of the world.

■ 12) What, Then, is the Moral of the Story? An Earthed Faith 12 (Ethics)

This volume will address questions around the relationship between Christian doctrine, Christian ethics, Christian spirituality, and Christian praxis—between the ultimate and the penultimate, between the indicative of God's grace and the imperative of ecological gratitude. Such relatedness has been there implicitly in all the other volumes but needs to be made explicit here. In dealing with climate change (for example), there is a need to find common moral ground with those standing in other religious traditions and with organizations in civil society. This has implications for all the relevant ethical categories—such as moral vision, virtues, duties, rights, responsibilities, values, middle axioms, action steps, and so on. For Christians, the question will be whether - and if so, how - such common moral ground is deeply rooted in the story of who God is and what God has done, is doing, and will be doing.

The Place of Creation in Christian Ecotheology – Some Shifts in the Story

Ernst M. Conradie[1] with Willie James Jennings[2]

■ The Series on "An Earthed Faith"

This is the third volume in the series on "An Earthed Faith: Telling the Story amid the 'Anthropocene.'" The aim of this series is to offer collaborative, constructive contributions to understanding the content and significance of central themes of the Christian faith from perspectives in Christian ecotheology, given the challenges associated with the so-called Anthropocene.

The assumption is that the Christian faith has a narrative shape and structure. It tells a story of who God is and what God has done, is doing, and might be doing within our world. The question, "Who is God?" is answered by telling a story (e.g., Deut 26:5-9). Clearly, there are different

---

1. Ernst M. Conradie is a senior professor in the Department of Religion and Theology at the University of the Western Cape in South Africa.

2. Willie James Jennings is an associate professor of Systematic Theology and Africana Studies at Yale University in New Haven, Connecticut in the United States of America. He is registered as a coresearcher at the University of the Western Cape for the project on "An Earthed Faith: Telling the Story amid the 'Anthropocene.'"

How to cite: Conradie, EM & Jennings, WJ 2024, "The Place of Creation in Christian Ecotheology – Some Shifts in the story", in EM Conradie & WJ Jennings (eds.), *The Place of Story and the Story of Place*, in An Earthed Faith: Telling the Story amid the "Anthropocene", vol. 3, AOSIS Books, Cape Town, pp. 1-24. https://doi.org/10.4102/aosis.2024.BK355.01

versions of this story, multiple perspectives on the story, and conflicting images and understandings of who God is, what God's identity and character may be, how we could possibly know that, and indeed, what being divine may mean. There are also many strands of narrative theology. What God may be doing is likewise contested. Yet somehow these stories and perspectives on Christian faith need to touch each other, unless there are different gods altogether, or unless God's actions are in conflict with each other, for example, suggesting that redemption is necessary to overcome the inadequacies of creaturely being. Throughout the ambiguous history of Christianity, it was far from easy to hold these together, and attempts at doing so readily became hegemonic. One therefore finds not one Christian tradition but many, not one faith but conflicting Christian faiths, not one history but contested his-stories, not a united but a fragmented Christianity. The definite article in "telling the story" is therefore clearly contested. Nevertheless, to use the indefinite article ("a Christian story") or the plural ("Christian stories"), without linking the various aspects of God's work to each other, runs the danger of allowing the various aspects of God's work to become disentangled and God's Triune identity to disintegrate.

Moreover, the contributions to this series are situated "amid the Anthropocene" with the recognition that a "business as usual" way of doing theology is no longer appropriate. Although the term "Anthropocene" is itself heavily contested,[3] it is at least clear that we live in a time where the balance between Earth subsystems has become disturbed and where (Western) Christianity stands accused by many to be one of the deepest causes of the underlying problem. How, then, should this story be told?

Volume 1 in the *An Earthed Faith* series offered a survey of the many strands of narrative theology[4] and then posed the question of how the Christian story of who God is and what God has done relates to the story of the universe. The second volume explored the emergence of ecotheology as a scholarly discourse, its global spread, and its fragmentation[5] in order

---

3. An increasing number of scholars in various disciplines raise questions about the use of this term because it obscures the particularity inherent in anthropogenic causation and in responsibility. By suggesting "humans" as the causal factor, the term may cloud over the reality that some humans are far more implicated than others in historic and contemporary greenhouse gas emissions, and that the lines of causation are highly racialized and class dependent. For the *An Earthed Faith* series, it was decided to use the term "Anthropocene" always in quotation marks to indicate the contestations over naming it as such. Doing ecotheology "amid" the "Anthropocene" is then not only a reference to disturbances in the Earth system but also to the dominant ways of interpreting it.

4. See the introductory essay by Ernst M. Conradie with Lai Pan-Chiu, entitled "On Setting the Scene for the Story to Begin."

5. See the introductory essay by Ernst M. Conradie and Cynthia Moe-Lobeda, entitled "Telling the Story *en Route*: On This Road (*hodos*) and Its Logic (*logos*)."

to raise questions around an appropriate method or methods of doing ecotheology.

■ The Question Posed in this Volume

The focus of this volume in the series is on creation theology. However this may be construed, God's work of creation forms an integral part of the story of who the Triune God is and what this God has done, is doing, and may be doing amid the "Anthropocene."

In each volume in the series, a single question is raised, which each of the contributors is asked to respond to from within their particular context and confessional tradition. Given this pattern, the third volume addresses the following question: "What difference does it make to the story of cosmic, planetary, human, and cultural evolution to re-describe this as the creative work of God's love? Inversely, what difference does it make to the story of God's love to describe it in evolutionary and geographic terms?" The ten contributors (two subsequently had to withdraw) were selected in order to optimize a diversity of positions in terms of geographical context, confessional traditions, and theological schools while also taking considerations of gender, race, age, and language into account.

Addressing this question requires theological reflection on the realities of place, such as land, geography, landscape, and a sense of being embodied, as well as evolution (cosmic, biological, hominid, and human) as the story of such a place and the roots of being embodied. Such reflection on the beginning is of course not situated "in the beginning" but entails a narrative reconstruction of the story where current interests, positions of power, and fears are necessarily at stake (the place where the story is being told). This is a contested space, indeed a "site of struggle," often dominated by issues of race rather than by grace. How, then, is this story to be told, given such a sense of place? It will not be possible to avoid questions around suffering, sin, evil, and the tragic (the theme of the fourth volume in the series), but the focus will be on why on Earth a loving God would deem this story to be "very good"—despite the prevalence of suffering, injustice, and oppression?

The aim of the series is ambitious, namely, to discern current paths in the state of the debate in Christian ecotheology, to identify emerging horizons in the field, and to take the debate forward through a set of constructive contributions to current discourse. While this introductory essay needs to paint in broad strokes in order to offer a proverbial bird's eye view of such current paths, the constructive contributions that follow will each need to speak from a particular place in order to take the debate forward.

In order to stimulate such further discussions, this introductory essay will offer some reflections on how the theme of creation has been approached in the specific context of Christian ecotheology. This is clearly by itself a story of many stories. It is also difficult to discern what may be included under the rubric of "ecotheology," especially because "ecology" has become a transversal in theological reflection on almost any other theme and in most discourses. The focus here is therefore on contributions that overtly recognize the need to address ecological concerns and would tend to use "ecotheology" as a form of self-description, allowing for some contestations on the term itself.

We propose that it is possible to identify at least seven shifting agendas in this story. In the mode of doing narrative theology, one might also refer to them as "chapters" in the story, but to avoid confusion, we will use the term "agendas." We do need to note that a clear chronological order is not implied, although we do find some sense of direction in this story. These shifting agendas overlap with each other in the work of many of the individual scholars mentioned below.

■ Agenda 1: The Self-critique of (Western) Male Theologies

In *God in Creation* (1985), Jürgen Moltmann describes the gradual but disastrous retreat from cosmology into personal faith in Western Christian theology.[6] This is a complex story in which the emergence of modern science played a crucial role, most notably in astronomy since the Copernican revolution (challenging an Earth-centered cosmology and prevailing notions of heaven), in geology (challenging prevailing notions of the age of the earth and therefore any literal reading of the Genesis creation narratives), and in evolutionary biology (challenging anthropocentric assumptions given human descent from other species). This increasingly challenged traditional forms of theism where God's ongoing interaction with the world could be readily assumed. In response, there were fluctuating movements toward deism (following Descartes and Newton), pantheism (following Spinoza, arguably also Hegel), atheism (following Marx, Nietzsche, Comte, and Freud), liberalism (following Schleiermacher), and fundamentalism, more specifically creationism.

Given such intellectual developments, Western Christian theology could find room for itself in interpreting faith in God as Creator existentially (following Luther's prompt that this means that God created me) or as a

---

6. See Moltmann, *God in Creation*, 36.

form of ongoing dependence (following Schleiermacher). Alternatively, it could focus on personal religious experience (pietism); on the message of the forgiveness of sin (guilt); on the church, its leadership structures, and its ministries to the faithful; on mission (following the voyages of exploration and colonial exploitation); and on matters of church and society (e.g., the social gospel), of ethics, or of interfaith dialogue. Such theological reflections could no longer do justice to the cosmic width of the Christian faith, did not overcome a Cartesian dualism with its emphasis on spirit instead of matter, and were vulnerable to a form of dominion theology that legitimized successive waves of Portuguese, Spanish, Dutch, Danish, French, British, and United States (US) imperialism, nineteenth-century notions of progress, the rise of capitalism in its various forms, industrialization, and the emergence of a consumer society.

An interest in creation theology was, of course, never completely lost. This is evident from the contrasting oeuvres of scholars such as Herman Bavinck (1854–1921), Karl Barth (1886–1968), Paul Tillich (1886–1965), Emil Brunner (1889–1966), Arnold van Ruler (1908–1970), Claus Westermann (1909–2000), and Gustaf Wingren (1910–2000), to name only some influential Lutheran and Reformed scholars. In such retrievals of creation theology, the debate hinged around the perceived dangers of natural theology. How can creation theology avoid a form of a theology based on nationhood, or on Nazi notions of "blood and soil," or on race as part of God's so-called creation ordinances (as in apartheid theology)? The danger is that any focus on what God created can serve as a legitimation of imperial, national, cultural, or class interests.

Nevertheless, the emergence of Christian ecotheology since the 1970s could be regarded as a critical response from within Western theologies to such a retreat from cosmology. At first, this was stimulated by a renewed interest in creation theology itself, often in conversation with insights emerging from the natural sciences. Soon, the focus shifted to anthropology and specifically to a theological understanding of the relationship between "man and nature," later revised to the place and role of humanity on (better: in) (the) earth, or the relationship between humans and nonhuman animals, typically inspired by and then itself inspiring an interest in environmental ethics. Such an interest in the theme of creation became evident across confessional divides, in multiple theological schools, and throughout the Western "world," including Scandinavian countries, Germany, the Netherlands, the United Kingdom (UK), the United States of America (USA), and among Orthodox churches in Eastern Europe.

A wide variety of positions soon developed, ranging from apologetic defenses of human dominion (or stewardship), revisionist, or more radical, nonanthropocentric positions.⁷ Put differently, some argued that the environmental crisis is a symptom of the failure to adhere to the command to exercise responsible stewardship, while others argued that the command itself (and the metaphor of stewardship that is employed) forms part of the problem insofar as it assumes human supremacy. Evangelicals typically welcome the notion of stewardship and seek to broaden this from financial responsibility to include environmental responsibility. Orthodox theologians are attracted to the notion of human priesthood, arguing that the priest is not only a mediator between God and humanity but also between God and the cosmos. However, in ecumenical reflections on church and society, such notions of stewardship became heavily criticized given the position of power and hierarchy that is assumed, and as God tends to be treated as an absentee landlord.⁸

■ Agenda 2: Widening the Scope of Ecotheology Beyond Creation Theology

Given the retreat from cosmology into personal faith as described above, there was an obvious need in Christian ecotheology to return to an adequate theology of creation, a theology of nature, to reconsider the need for, or the inevitability of, natural theology, to rethink the relationship between humans and other creatures, and to draw implications for environmental responsibility. However, once the renewed interest in creation theology was

---

7. It is not possible to offer a comprehensive survey of the many contributions to Christian ecotheology where such a self-critique of Western (male) creation theology became evident and where there is some reference to typical concerns in Christian ecotheology. Suffice it to mention publications in English by Western authors such as Thomas Berry, John Black, John B. Cobb, Thomas Sieger Derr, Calvin DeWitt, Matthew Fox, Colin Gunton, Douglas John Hall, Philip Hefner, George Hendry, Jürgen Moltmann, H. Paul Santmire, Joseph Sittler, Michael Welker, Loren Wilkinson, and surely many others. In German, one would also need to mention important early contributions by Günter Altner, Gerard Liedke, and Christian Link. One may also mention contributions by Orthodox theologians such as Paulos Gregorios (Paul Varghese), John Zizioulas, and others. Such a listing of scholars contributing on creation theology in the field of ecotheology cannot be comprehensive and serves here merely as an open invitation to add other important contributions. Despite the obvious conflicting variety of approaches and contrasting positions on anthropocentrism, one may find a common intuition that Western (male) theologies have become distorted as a result of a retreat from cosmology into personal faith, as a result of a Cartesian dualism and an arrogant form of dominion theology legitimizing forms of human domination, including imperial and colonial conquest.

The following influential books may be mentioned in the order of date of first publication: John Black, *The Dominion of Man* (1970); Günter Altner, *Schöpfung am Abgrund* (1974) and *Zwischen Natur und Menschengeschicht* (1975); Gerard Liedke, *Im Bauch des Fisches* (1979); Matthew Fox, *Original Blessing* (1983); Jürgen Moltmann, *God in Creation* (1985); Paul Gregorios, *The Human Presence* (1987); Christian Link, *Schöpfung Band I* and *Band II* (1991); and Phil Hefner, *The Human Factor* (1993).

8. The emphasis on stewardship may be found in the oeuvres of (all male!) scholars such as Sam Berry, Thomas Sieger Derr, Calvin DeWitt, and especially Douglas John Hall, *The Steward*. For a critical discussion see Conradie, *Christianity and Earthkeeping*.

established, an inverse trend became evident, namely, to reconnect the theme of creation with other aspects of the Christian faith.

This may be understood in terms of the distinction between God's act of creation (*creatio*) and the creation that results from such work (*creatura*). Once the focus is on *creatura* (whether as human, animal, ecosystem, land, atmosphere, biosphere, hydrosphere, planet, or cosmos), that is, on that which is material, bodily, and earthly, there emerges an obvious theological need to bring themes such as sin, providence, salvation, church, the sacraments, mission, and eschatological consummation into play as well. Moreover, each creature may be considered from distinct, that is, Patrological, Christological, and Pneumatological perspectives, each yielding different but complementary insights. This is sometimes described as "Trinitarian spreading," suggesting an interest not only in the work of God but also in the person of God.

Such a widened doctrinal scope became evident in contributions to ecotheology at least since 1990. These perspectives will be discussed in subsequent volumes within this series. This series is itself born from such a recognition that ecotheology cannot be reduced to creation theology, anthropology, or environmental ethics, or else that can easily lead to theological self-marginalization.

What should be noted here is that creation theology itself could no longer be regarded as standing on its own but had to be connected with such other aspects.[9] One finds such connections in titles with an "and," such as creation and salvation, creation and eschaton, creation and church, Christ and creation, Spirit and creation, liturgy and creation, mission and creation, and so forth. Indeed, such a need may by now be taken for granted. However, as two major volumes produced within the context of the "Christian Faith and the Earth" project demonstrated, to hold together God's work of creation *and* of salvation is easier said than done.[10] There are

---

9. Such issues need not be explored here in any further detail. One may say that such a doctrinally comprehensive approach is evident in the whole oeuvres of most of the leading Western-trained scholars in the field of Christian ecotheology, including Sigurd Bergmann, Leonardo Boff, Steven Bouma-Prediger, Celia Deane-Drummond, Denis Edwards, Matthew Fox, Elizabeth Johnson, Catherine Keller, Sallie McFague, Jürgen Moltmann, James Nash, Michael S. Northcott, Rosemary Radford Ruether, Paul Santmire, Peter Manley Scott and Mark I. Wallace. Although such authors come from very, very different theological traditions, they share a common intuition that addressing ecological concerns requires a doctrinally comprehensive approach. An adequate creation theology is needed but cannot stand on its own. Beyond such a common intuition, confessional and other differences lead to very different ways of organizing the material. See especially the one-volume overviews of ecotheology by Paul Santmire, *The Travail of Nature* (1985); James Nash, *Loving Nature* (1991); Sallie McFague, *The Body of God* (1993); James Gustafson, *A Sense of the Divine* (1994); Denis Edwards, *Ecology at the Heart of Faith* (2006); Steven Bouma-Prediger, *For the Beauty of the Earth* (2007); and Celia Deane-Drummond, *Eco-Theology* (2008).

10. See the two volumes entitled *Creation and Salvation* (2011, 2012) produced within the context of the "Christian Faith and the Earth" project, both edited by Ernst M. Conradie.

major confessional differences on whether God's work of salvation restores, elevates, divinizes, replaces, or recycles the fallen creation. There are also major theological differences on whether creation and salvation can be fused as two dimensions of the same ongoing process—with the twin dangers that creation can be regarded as God's only work (deism) or where salvation is regarded as necessary to overcome the inadequacies of God's work of creation (given natural evil).[11]

■ Agenda 3: Creation, the Natural Sciences, and Ecological Concerns

In Western ecotheology, one also finds a different, partially overlapping trajectory. Once the interest in the theme of creation is firmly re-established, it is also possible to focus on conversations with a wide range of natural sciences to gain a better understanding of what is material, earthly, and bodily. The focus could be on astrophysics, quantum cosmology, thermodynamics, chemistry, the atmospheric sciences, geology, evolutionary biology, animal ethology, or on "ecology" in the narrower sense of studying specific ecosystems or bioregions, or, in a much wider sense, on Earth system science. The interest could be in understanding the significance of God's work of creation and ongoing creation, or, alternatively, on a more detailed description of environmental impact. Sometimes, scholars working in such fields have prior training in such disciplines or engage in interdisciplinary or multidisciplinary projects, but in one way or another, there is then a longer route to return to theological and ethical reflection via such conversations with the natural sciences. Some also recognize the need to bring the social sciences (including economics), the arts, and the humanities into play, but the focus then tends to shift to the human creature only. Often, this leads to a focus on ethical concerns, for example, on ecojustice at multiple levels, while the doctrine of creation is not addressed directly.

Theological discourse holding together the themes of God's creation, the natural sciences, and ecological concerns does not necessarily employ "Christian ecotheology" as a form of self-description but is clearly relevant here as well.[12]

---

11. For a detailed discussion of these problems, see Conradie, *Saving the Earth?;* also *The Earth in God's Economy.*

12. An early marker in this regard may be the World Council of Churches (WCC) conference on *Faith, Science and the Future* hosted in Boston in 1979. See the pre-conference and post-conference volumes edited by Paul Abrecht.

Not surprisingly, one finds widely diverging approaches in such conversations between theology and the natural sciences on environmental concerns. Only the briefest of sketches will have to suffice here, without detailed references, as these could fill volumes.

Firstly, one finds sometimes overt and sometimes implicit references to ecological concerns in the work of leading scholars in the fields of "science and theology" or "science and religion," including the quite considerable oeuvres of Ian Barbour, Philip Clayton, Philip Hefner, Alister McGrath, Arthur Peacocke, Ted Peters, John Polkinghorne, Nancey Murphy, Holmes Rolston III, Robert John Russell, Christopher Southgate, William R. Stoeger, and many others who engage with creation theology in one way or another.

Secondly, the legacy of Teilhard de Chardin and his synthesis of Christianity and evolutionary history remains evident in the work of Thomas Berry, Heather Eaton, Anne Marie Dalton, Brian Swimme and Mary Evelyn Tucker. For Berry (who often described himself as a "geologian"), especially, there is a need to connect the insights from astrophysics, the geological sciences, and evolutionary biology in order to reconstruct the "story of the universe."[13] For Berry and his followers, there is a clear ecological moral to the universe story.[14]

Thirdly, the legacy of Teilhard is often related to process theology, again with addressing ecological concerns in mind. This is, for example, evident in the work of John F. Haught and Jay McDaniel, but the influence of process theology is also evident in many of the scholars mentioned elsewhere in this introductory essay, including Ian Barbour, Philip Clayton, Catherine Keller, Thomas Jay Oord, and especially John B. Cobb. In the oeuvre of John F. Haught, such insights are brought together under the rubric of the "promise of nature" in order to address a sense of cosmic homelessness that is exacerbated by ecological destruction.[15]

Fourthly, a number of Christian theologians are attracted to the Gaia hypothesis, that is, the notion that the earth may be regarded as a single, self-regulating organism (i.e., a living planet), as proposed by James Lovelock, Lynn Margulis and others on the basis of insights in planetary systems. This is developed especially by Anne Primavesi throughout her oeuvre, but allusions to the Gaia theory may also be found in theologians as far apart as Jürgen Moltmann and Rosemary Radford Ruether.

---

13. See Berry and Swimme, *The Universe Story*.

14. See Swimme and Tucker, *Journey of the Universe*. See also the critique of such approaches by Lisa Sideris in *Consecrating Science*.

15. See Haught, *The Promise of Nature*.

Fifthly, there is considerable interest in Christian ecotheology in the problem of "natural evil." There is consensus that social evil greatly exacerbates such "natural evil" and that the environmental crisis is a manifestation of that. However, the "groaning of creation" is rooted in problems associated with the limited life cycle of cells, degeneration in multicellular organisms, vulnerability, pain, suffering and anxiety among sentient animals, transience, mortality, natural nonselection, extinction, and even gravity itself causing injuries as a result of falling down. Given such natural suffering, there is a need to reinterpret notions of human finitude, sin, salvation, and the eschaton. This nexus of problems is widely discussed in literature on science and theology and is related to ecological concerns most clearly in the oeuvre of Christopher Southgate.[16] In response to such a focus on natural sources of suffering, others have insisted that this should not draw attention away from anthropogenic causes of ecological destruction, especially given the role of colonization, white supremacy, and patriarchy.[17]

Finally, creation theology in the context of ecotheology could also be concerned more narrowly with theological reflection on animals other than humans. The focus could be on issues of biodiversity, the extinction of species and nature conservation. Or it could focus on animal ethology from an evolutionary perspective. Alternatively, the concern may be ethical, especially with reference to the plight of farm animals in industrialized agriculture, the injustices embedded in the food economy as a whole, or against household pets.[18] As will become evident in the discussion to follow, such engagement with the natural sciences that is typical of many contemporary Western creation theologies may be contrasted with attempts to retrieve an Indigenous worldview in many other forms of Christian ecotheology.

---

16. See Southgate, *The Groaning of Creation*; also *Theology in a Suffering World*. The problems posed by natural suffering will also be addressed in Volume 4 of this series, namely, on the notion of God's providence and the theodicy problem that this raises.

17. See Conradie, "On Social Evil and Natural Evil."

18. In addition to earlier contributions, especially by Bernd Janowski and Gerard Liedke, animal theology flourishes in the work of British scholars such as Celia Deane-Drummond, David Grumett, Andrew Linzey, Rachel Muers, and, more recently, especially David L. Clough, *On Animals Volume 1* (2012) *and Volume 2* (2019). Such themes continue to attract considerable attention in Germany, for example, through a recent Evangelische Kirche in Deutschland (EKD) study. See EKD, *Livestock and Fellow Creatures!*

Agenda 4: The Shift to (Western) Ecofeminist Theologies of Creation

Western feminist theologies are arguably characterized by a dual critique, namely, a Christian critique of patriarchy and therefore of domination in the name of differences of gender and sexual orientation, as well as a feminist critique of Christianity, its history, sacred texts, institutions, practices, missions, ministries, and theologies. There is no need to reiterate such critiques here except to note that Christian ecotheology is also characterized by such a dual critique—an ecological critique of Christianity and a Christian critique of ecological destruction.

Most feminist theologians recognize the need for a retrieval of some form of creation theology precisely in order to overcome the way creation theologies are all too often employed in the service of or to explicitly legitimize patriarchy. The feminist critique of interlocking dualisms is found in most, if not all, such creation theologies. There is an appreciation for what is immanent, material, bodily, and earthly, polemically directed against a (male?) emphasis on what is transcendent, ideal, spiritual, and heavenly. Such feminist creation theologies are therefore drawn toward forms of panentheism in order to overcome the dual challenges of deism (God's utter transcendence) and pantheism (God's complete immanence). Some feminist scholars, for example, Elizabeth Johnson, Sallie McFague and Susan Rakoczy, remain deeply Trinitarian in their thinking, while others would tend to agree with Mary Daly that if God is male, then the male is god—and therefore explore ways of naming the divine in other religious traditions and forms of spirituality.

Female experiences of being embodied are highlighted and explored over and against male experiences, for example, with reference to birthing, mothering, and nurturing. This may be a risky move given the association of female bodies with "Mother Earth," as this may merely reinforce the set of interlocking dualisms that also include the divides between human and animal, reason and passion, aggression and vulnerability, curing and caring, conflict and cooperation, and God and the world. Nevertheless, the intuition is that female experiences of being embodied can provide a lens for understanding both ecological destruction (the "rape" of "Mother Earth") and ecological integrity (symbiosis). It comes as no surprise, then, that most scholars who use the adjective "feminist" as a form of self-description would nowadays also welcome an ecofeminist approach. In their constructive contributions, feminist (creation) theologies typically emphasize relatedness and explore a variety of metaphors to express relationships based on mutual respect, mutual care, and reciprocity, also in human interactions with otherkind.[19]

■ Agenda 5: Liberation, Black, Womanist, and Subaltern Creation Theologies[20]

Creation theology is all too often produced by a royal or imperial elite to legitimize their occupation of land. This danger was already prevalent in biblical times with the conquest of Canaan, the extension of Solomon's empire, and nationalist claims for Zion to be God's abode. Accordingly, if the earth is said to be the Lord's (Ps 24:1), then the land surely belongs to feudal landlords as God's representatives on earth!? This danger has plagued the history of Christianity ever since Emperor Constantine declared Christianity to be the state religion with the Edict of Milan in 313. In this form, creation theology could provide a justification for the Holy Roman Empire and for the Crusades, while Eastern and Oriental Orthodoxy could provide religious legitimacy for various nation-states, including Ethiopia, Greece, and Russia.

Creation theology also provided the basis for the 15th-century voyages of exploration and exploitation and for the subsequent imperial and colonial conquest of wave upon wave of successive empires, whether Portuguese, Spanish, Dutch, Danish, French, Belgian, German, British, or US.[21] The bull issued by Pope Nicholas V in 1452 shows how creation theology could be used to justify and sanctify colonial plundering:

---

19. Such a shift towards Western ecofeminist theologies of creation may be found in numerous landmark publications across the North Atlantic region. It may again suffice here to mention the work of scholars such as Mary Daly, Heather Eaton, Mary Grey, Catharina Halkes, Grace Jantzen, Elizabeth Johnson, Catherine Keller, Sallie McFague, Rosemary Radford Ruether, and Dorothee Sölle, where ecofeminism, Christian ecotheology, and a focus on creation theology are held together. Further references can surely be multiplied, and this comes again with an open invitation for others to add further examples.

The following influential books published in English between 1970 and 2000 may be mentioned in the order of date of first publication: *New Woman, New Earth: Sexist Ideologies and Human Liberation* (1975) by Rosemary Radford Ruether; *Gyn/Ecology: The Metaethics of Radical Feminism* (1978) by Mary Daly; *God's World, God's Body* (1984) by Grace Jantzen; *To Work and to Love: A Theology of Creation* (1984) co-authored by Dorothee Sölle and Shirley Cloyes; *Redeeming the Dream: Feminism, Redemption and Christian Tradition* (1989) by Mary Grey; *New Creation: Christian Feminism and the Renewal of the Earth* (1991) by Catharina Halkes (Dutch version 1989); *Gaia & God: An Ecofeminist Theology of Earth Healing* (1992) by Rosemary Radford Ruether; *The Body of God: An Ecological Theology* (1993) by Sallie McFague; *On Earth as in Heaven: A Liberation Spirituality of Sharing* (1993) by Dorothee Sölle; *Women, Earth, and Creator Spirit* (1993) by Elizabeth Johnson.

20. As an aside, one may observe that ecofeminist contributions in contexts other than the North Atlantic, including Latin America, Africa, South and East Asia, are characterized by a collaborative spirit so that publications of such a kind are more typically found in edited volumes than in monographs. See, for example, Hallman, *Ecotheology: Voices from South and North* (1994), Ruether, *Women Healing Earth* (1997); Kim and Koster, *Planetary Solidarity* (2017).

21. For a discussion, see Jennings, "Reframing the World."

[...] to invade, search out, capture, vanquish, and subdue all Saracens and pagans whatsoever, and other enemies of Christ wheresoever placed, and the kingdoms, dukedoms, principalities, dominions, possessions, and all movable and immovable goods whatsoever held and possessed by them and to reduce their persons to perpetual slavery, and to apply and appropriate to himself and his successors the kingdoms, dukedoms, counties, principalities, dominions, possessions, and goods, and to convert them to his and their use and profit.[22]

Creation theology in this mold is evident at its worst in German views on "blood and soil," in apartheid theology's justification of racial segregation on the basis of creation ordinances, and in American assumptions about "our manifest destiny" where the "our" is built upon slavery and White supremacy, where Black lives clearly do *not* matter. Given this situation, the interest in the theme of creation in Western theologies, especially given the association with science and technological power, but also in the context of ecotheology, needs to be subjected to a hermeneutics of suspicion. In the words of Latin American theologian Vítor Westhelle, where landlords looking over their estates see the beauty of God's creation, peasants and the landless can only see gates and fences designed to keep them out.[23] The poor, oppressed, and marginalized are rarely interested in theologies of creation that function as theological justification for current social, political, and economic arrangements. Rather, the theologians and ethicists who write from and on behalf of such communities had in the past focused on creation care through the lens of liberation from oppression and the brutality associated with the West. Yet their work highlights the danger of forming theologies of creation that help to legitimate new forms of oppression and marginalization. However, in recent years, more theologians of the Two-Thirds World have been focusing especially on the links between the colonial operations that exploited and continue to exploit and that destroyed and continue to destroy humans and those that do the same to animals and the environment.[24]

---

22. See Indigenous Values Initiative, "Dum Diversas," Doctrine of Discovery, July 23, 2018, https://doctrineofdiscovery.org/dum-diversas/.

23. See Westhelle, "Creation Motifs in the Search for a Vital Space."

24. See Boff, *Cry of the Earth, Cry of the Poor*, 104–14, as well as *Cry of the Earth, Cry of the Poor*, 81–85. The integral relatedness of poverty and ecological destruction became widely recognized following an edition of *Concilium* entitled *Ecology and Poverty: Cry of the Earth, Cry of the Poor* (1995), edited by Leonardo Boff and Virgilio Elizondo. This connection was further developed in Boff's *Ecology and Liberation* (1995) and in his *Cry of the Earth, Cry of the Poor* (1997) and gained papal prominence through Pope Francis's encyclical *Laudato Si'*. Boff insists that liberation theology and ecotheology are partners, not rivals. He does offer a distinct creation theology by following Teilhard's vision of cosmogenesis to diagnose the ultimate root of the ecological crisis as a disruption of universal connectedness. Such insights are by now widely found among Latin American liberation theologians such as Ivone Gebara, Guillermo Kerber, Vítor Westhelle, and many others. In his recent *An Ecological Theology of Liberation* (2019), Daniel Castillo retrieves Gustavo Gutiérrez's vision of liberation to interpret Pope Francis's notion of integral ecology. He includes a chapter on reading Genesis in which he emphasizes the tension between de-creation and re-creation but says rather little about God's work of creation beyond its peaceable way of bringing things forth.

It became abundantly clear that the economic forces that lead to poverty and inequality are the same forces leading to ecological destruction, namely industrialized capitalism, especially in its neoliberal, globalized forms. Such forces build upon imperial and colonial conquest and are expressed through neocolonial control over production, trade, finances, and markets. Its logic is one of sustained economic growth. As a result, the land itself is crying for justice.[25]

Two qualifications must be added. One is that industrialized socialism also assumes economic growth in the production of wealth, although it differs on the proper distribution of such wealth. The second is that the emphasis on the production of wealth should be complemented by a concern not only over the ownership of wealth but also over the consumption of such wealth. This led to an interest in the impact of consumerism, not only among the affluent ("affluenza") but also its devastating impact on the aspirations of the poor.[26]

The relatedness of economic and ecological injustices was very soon extended to recognize an underlying logic of domination in the name of difference, not only in terms of class but also in terms of race, gender, sexual orientation, caste, language, and culture. This insight is by now widespread in a range of self-consciously contextual theologies. "Ecology" is thus regarded as a transversal and a dimension of any and all other discourses. Indeed, everything is now ecological, even if "ecology" is not necessarily everything. While such transversals typically focus on the human creature, some would also acknowledge that the same logic of domination is also at work in the name of differences between species. This is then expressed in concerns over the plight of animals (the "new poor"), indigenous forests, and the degradation of the land itself.

On this basis, there is an emerging consensus that environmental justice cannot be separated from other forms of justice and vice versa. This is particularly well captured in a landmark paper by James Cone entitled "Whose Earth is it Anyway?" at a conference on *Ecumenical Earth* at Union Theological Seminary in New York City in 1998. Cone challenges the ecological movement to critique itself through a radical and ongoing engagement with racism and likewise challenges the Black liberation movement to take a critical look at itself through the lens of the ecological movement.[27]

---

25. See the statement endorsed by church leaders in South Africa before the World Summit on Sustainable Development in Johannesburg (2002), entitled *The Land is Crying for Justice*.

26. For one overview of such debates, see Conradie, *Christianity and a Critique of Consumerism*.

27. Cone, "Whose Earth Is It Anyway?"

This insight was built upon and expanded through the theological and ethical insights of womanist theologians and ethicists like Katie Geneva Cannon, Karen Baker-Fletcher, M. Shawn Copeland, and Melanie Harris. Each, in different ways, articulated the inseparable connection between environmental justice and the thriving of Black women and other women of color in the Global South.[28] In her widely acclaimed *Longing after Running Water* (1999), the Brazilian ecofeminist theologian Ivone Gebara explores the connections between ecofeminism and liberation. Significantly, she does not include a chapter on creation theology and, in the prologue, explains that her personal experience is "totally urban."[29]

On this basis, the term ecojustice is widely used in ecumenical discourse to capture the need for a comprehensive sense of justice that can respond to economic injustice, ecological degradation *and* the interplay between them. It builds upon the widespread recognition that the English words ecology, economy, and ecumenical share the same etymological root in the Greek *oikos* [household]. Accordingly, ecology describes the underlying logic [*logos*] of the household, economy circumscribes the rules [*nomoi*] for the management of the household, while the "whole inhabited world" [*oikoumene*] refers to the (human) inhabitation of the household.

Subsequently, this core insight is addressed in diverging ways in African, Black, feminist, LGBTQIA+, liberation, Minjung, Mujerista, subaltern, and womanist theologies, depending on the particular struggle against domination in the name of difference. This is not necessarily framed as a form of creation theology, but it would also miss the point not to recognize the concern over that which is material, bodily, and earthly over a sense of place and habitat.[30]

---

28. See Baker-Fletcher, *Sisters of Dust, Sisters of Spirit: Womanist Wordings on God and Creation* (1998). Also see Emilie M. Townes et al., eds., *Walking Through the Valley: Womanist Explorations in the Spirit of Katie Geneva Cannon* (2022) and Melanie Harris, *EcoWomanism* (2017).

29. Gebara, *Longing for Running Water*, v.

30. This may be illustrated by the case of South African Black theology. At its core, South African Black theology is a theology of liberation and thus a form of soteriology, not creation theology. Clearly influenced by Latin American liberation theology, its sources of inspiration include the civil rights movement in the USA, North American Black theology, the Black consciousness movement led by Steve Biko, and of course the South African political struggle for liberation from imperialism, colonialism, and apartheid. Its notion of liberation is necessarily multilayered, including political, military, economic, social, and intellectual dimensions. As a theology, it portrays God as a Liberator, on the side of the poor and oppressed, the Black working class, the marginalized. Where such liberation is not forthcoming, this requires social analysis and critique, easily turning into a theodicy: why does God allow such suffering to continue unabated?

Black theology is closely aligned but not to be equated with liberation theology. This is suggested by adding the adjective "Black" to Black liberation. Its core metaphor is rooted in African humanism, affirming the dignity and, indeed, the pride of colonized people: Black is beautiful. Its emphasis is not only on liberation from oppression; it also resists replacing the colonial masters with new forms of oppression. This early decolonial critique is possible through a retrieval of the category of Blackness. As such, it is a form of creation theology, even if not in any overt form. To be sure, Blackness was not reduced to skin pigmentation,

■ Agenda 6: The Retrieval of Indigenous Theologies of Creation Globally

While liberation, Black, womanist, and subaltern theologies typically have an implicit more than explicit focus on creation, the same does not apply to Indigenous theologies. This is clearly related to a sense of belonging to ancestral land and to a retrieval of Indigenous ecological wisdom. The critique of colonial conquest remains firmly entrenched, however. The earth may be the Lord's, but it clearly does not belong to colonial landlords (or to boards of directors of agribusinesses).

Such Indigenous creation theologies may be found around the world. The stories may be different, but they respond to an underlying pattern that is unmistakable, namely a process of imperial and colonial military conquest followed by the occupation of the land by foreign powers, exposure to new diseases and pandemics, a contraction of the land previously occupied, leading to the marginalization of Indigenous peoples, and a subduing of their cultures through the economic, social, and educational dominance of the ruling class. By contrast, liberation, Black, womanist, and subaltern theologies follow upon the slave trade and oppression on the basis of differences of class and caste. In the African context, a retrieval of Indigenous ecological wisdom in response to a history of colonization, deforestation, and the destruction of ancestral land is common to many contributions to ecotheology. Such wisdom is often built upon African creation narratives in juxtaposition with the Christian story of God's work.[31]

---

hair, or the size and shape of noses (as Desmond Tutu would often joke). Black consciousness is a form of *consciousness*, a way of thinking and being on the side of the oppressed. It could therefore be inclusive by resisting the race classification under apartheid. Yet *Black* consciousness allowed for a recognition of what is concrete, material (the alienated means of production), and bodily (displacement, forced removals, suffering, and torture). It also included a gendered component, although this became more prominent in theological reflection only after 1994.

Black theology as creation theology did not invite a romantic retrieval of a precolonial past, given its focus on the class struggle. It hardly addressed ecological concerns before 1994. It is at best a truncated creation theology with a heavy emphasis on anthropology. This anthropological focus is evident from the writings of leading exponents such as Allan Boesak, Manas Buthelezi, Frank Chikane, Bonganjalo Goba, Simon Maimela, Takatso Mofokeng, Itumeleng Mosala, Mokgethi Motlhabi, Buti Tlhagale and Desmond Tutu. The creation-centred emphasis of Gabriel Setiloane may serve as a significant counter-example. The core theme is that if all human beings are created in the image of God, this applies to the poor, the oppressed, the marginalized, and the tortured too—even to oppressors (as Desmond Tutu reiterated). Its creation theology is therefore inclusive, even if its focus tended to remain anthropocentric. This interpretation of Black theology as a form of creation theology draws almost verbatim on Conradie, "Doing Creation Theology in the South African Context."

31. One may mention the early work of scholars such as Emmanuel Asante, Marthinus Daneel (e.g. *African Earthkeepers*), Jesse Mugambi (e.g. *God, Humanity and Nature*), Gabriel Setiloane, and Harvey Sindima and more recently the contributions of Robert Agyarko, Emmanuel Amin, Ezra Chitando, Samson Gitau, (e.g. *The Environmental Crisis*), Ben-Willie Kwaku Golo, Kapya Kaoma (e.g. *God's Family, God's Earth*), Chammah

While some African contributions retrieve Indigenous ecological wisdom in African culture, others seek to come to terms with the rise of neo-Pentecostalism and the popularity of the prosperity gospel. Prophetic and liberation theologies focus on issues of international debt, the environmental impact of structural adjustment programs, the long-term impact of extractive colonialism, climate justice, and climate adaptation. Contributions to ecotheology from the Pacific region are becoming increasingly significant, not least given experiences of the impact of rising sea levels. Such voices are prominent in the lobbying of Small Island Developing States (SIDS) at the various Conferences of the Parties, at consultations of the World Council of Churches, and through postgraduate studies by church leaders. In addition to prophetic critique, the constructive contributions typically draw on Indigenous wisdom embedded in a theology of creation where relationality is emphasized.[32]

Such contributions from around the world may not be structured in the traditional form of creation theologies emerging from the North Atlantic region, but there can be no doubt about the focus on the well-being of the land and all its living creatures. Often, biblical narratives are interweaved with Indigenous stories. A striking feature of such contributions is the nonanthropocentric approach that is often adopted, in the sense that humans belong to the land and not the land to humans. The starting point for creation theology is therefore inhabitation. There is a sense of being embodied and that human bodies are intimately connected to those of others. All beings are regarded as animate so that hierarchies and clear boundaries between minerals, plants, animals, and humans tend to become dissolved.

■ Agenda 7: Ecumenical Efforts Toward Telling the Story Together

The bird's eye view of how the story of creation (and of creation theology) has been told in the context of Christian ecotheology, as outlined, shows a

---

Kaunda, and Teddy Sakupapa in this regard. See also Mugambi, and Vähäkangas, *Christian Theology and Environmental Responsibility* (2001).

Ecotheology flourishes especially in the context of the Circle for Concerned African Women Theologians, as is illustrated by an early volume entitled *Groaning in Faith: African Women in the Household of God*, edited by Musimbi Kanyoro and Nyambura Njoroge (1996). The work of Mercy Oduyoye remains foundational, while the more recent contributions by Sophia Chirongoma, Musa Dube, Mary Getui, Eunice Kamaara, Susan Kilonzo, Loreen Maseno, Fulata Lusungu Moyo, Isabel Mukonyora, Kuzipa Nalwamba, Isabel Phiri, Lilian Siwila, Gloriose Umuziranenge, and others are noteworthy.

32. An early example is the PhD thesis by Ama'amalele Tofaeono (Samoa) entitled *Eco-Theology: Aiga—The Household of Life* (2000). Subsequent contributions may be found in the work of Faafetai Aiava (Samoa), Cliff Bird (Solomon Islands), Aisake Casimira, Jione Havea, Tafue Lusama, Maina Talia, and Upolu Lumā. Vaai (Samoa and Fiji). There is a rapidly expanding corpus of such literature. See, e.g. Bhagwan et al. eds. *From the Deep*.

conflicting plurality of positions and approaches. The place where each story is told matters, but each place also has a contested story. Often, the stories are about fiercely contested spaces. If a particular place exists in the singular, the stories are necessarily in the plural. These stories cannot be integrated with one another in ecumenical harmony without running roughshod over a history of imperial and colonial conquest, often in the name of Christianity. Christianity, too, exists in the plural: there are multiple Christian traditions and contested Christianities.

The ecumenical movement, in its classic and contemporary forms, is built on the intuition that the plural, such a conflicting plurality, cannot have the last word. One reason for that is that contestation may be over the same place. Moreover, places are related to each other within the one *oikoumene*, the whole inhabited world. Our understanding of that world may keep changing, and the world itself is subject to change through cosmic, biological, and cultural evolution, but the earth as a planet exists in the singular. This is confirmed by earth system science. There may be multiple subsystems, and how these interact with each other is incredibly complex, but the earth system exists in the singular. The term "Anthropocene" suggests that the noosphere affects the atmosphere, that changes in the atmosphere will affect the biosphere and will become embedded in the lithosphere for millions of years. At the same time, the ecumenical critique of neoliberal globalization (resisting a new form of empire-building) and the vehement contestations over naming the "Anthropocene" as such suggest that any talk about the *oikoumene* in the singular is fraught with the dangers of imperialism.[33]

Another reason for the singular is more strictly theological. Even if one accepts conflicting Christianities and creation stories in the plural, does that imply that there are also different creators, not only different notions of the Creator? Is the world perhaps then the outcome of the conflict between different gods altogether? Are these gods more or less on par with each other?

The Nicene–Constantinopolitan Creed (381), the reference point for the classic ecumenical movement, did not speak of the Creator in the singular. It confessed faith in God as Creator in a triune way. The "one God," "the Father, the Almighty" [Πατέρα, παντοκράτορα; Latin *Patrem omnipotentem*] is the "maker of heaven and earth"; through the "one Lord, Jesus Christ," "all things were made"; while the "Holy Spirit, the Lord," is "the giver of life." Three ways, then, to confess the "one God" [ἕνα Θεόν] and what this God made or "composed", namely "heaven and earth, all that is, seen and unseen" [οὐρανοῦ καὶ γῆς, ὁρατῶν τε πάντων καὶ ἀοράτων], "all things"

---

33. See Vaai, "We are Earth."

[τὰ πάντα], and "life" [ζῳοποιόν]. Note that the Greek verbs used are ποιητὴν, ἐγένετο, and -ποιόν, although these are translated into Latin as *factorem*, *facta sunt*, and *vivificatorem*. In short, there is an ecumenical need to hold these three ways together in order to keep speaking of "all things," what is visible or not, living or not.

The contemporary ecumenical movement continues to recognize this need to speak of the whole inhabited world, to hold together the wide variety of Christian stories about God as the Triune Creator.[34] Christian ecotheology cannot be understood apart from, and has been at the forefront of, such ecumenical efforts. This is exemplified by the global spread of Christian ecotheology in the full spectrum of geographical contexts, confessional traditions, schools of theology, language traditions, and cultural contexts. At its best, it has addressed the global divides between North and South, East and West, so-called developed and so-called developing countries, centers, and margins. However, contemporary Christian ecotheology itself also reflects such divides.[35]

Such global ecumenical dialogue across the divides remains vital, but it is much harder than is often assumed. That applies especially to North–South and East–West dialogues where issues of language, worldviews (scientific or Indigenous), manifestations of religious plurality, and socioeconomic power inhibit respectful but also mutually critical conversation. Even where there is North–South dialogue, it is often between like-minded scholars who agree to collaborate so that the harder conversations do not actually take place. It matters where initiatives come

---

34. This is, for example, evident in various mottos derived from assemblies of the WCC. The Nairobi Assembly (1975) spoke of a "Just, Participatory and Sustainable Society." The Vancouver Assembly (1983) introduced the conciliar process towards "Justice, Peace and the Integrity of Creation" (JPIC). The theme of the Canberra Assembly (1991) was the prayer "Come, Holy Spirit, renew the whole creation." The Harare Assembly (1998) had "Turn to God, Rejoice in Hope," and the Porto Alegre Assembly (2006) followed a similar line, "God, in Your Grace, Transform the World," while the Busan Assembly (2013) had "God of Life, Lead us to Justice and Peace." The theme of the Karlsruhe Assembly (2022), "Christ's Love Moves the World to Reconciliation and Unity," likewise speaks of the world in the singular. Note the use of God, Christ, and Holy Spirit as Triune ways of seeing the world.

35. In a landmark recent volume entitled *Decolonizing Ecotheology* (2021), edited by Lily Mendoza and George Zachariah, *Indigenous and Subaltern Challenges* (as the subtitle has it) to what is described as "mainstream ecotheology" and its dominant forms of creation theology are explored. The authors resist the colonization of the commons and retrieve a notion of commonwealth, namely "the community of creation that births, sustains and nurtures the movement of life—the land, water, forests, seeds, biodiversity, atmosphere and the commoners who live in communion with these commons." The collaborative "Christian Faith and the Earth" project (2007–2014), one of the main stimuli behind this series, may serve as an example of efforts to do theology ecumenically, across the many global divides. It involved more than 100 scholars from six continents to encourage further debate in ecotheology across geographical contexts and confessional traditions. It yielded several edited volumes, including one from its culminating conference entitled *Christian Faith and the Earth: Current Paths and Emerging Horizons in Ecotheology* (2014). This series on *An Earthed Faith* clearly builds upon such earlier initiatives.

from, what the language of communication is, what the countries of origin and current location of contributors are, who the dominant conversation partners are and in which disciplines they are found, where publishers and their perceived markets are located, how books are priced, and whether or not they are open access. Within ecotheology itself, it seems that evangelical proponents of responsible stewardship in the consumer classes or Orthodox proponents of priesthood seldom talk to climate change activists in subaltern movements. There are diverging positions on how to address religious plurality theologically, but these are seldom addressed. As far as creation theology itself is concerned (i.e., not only environmental ethics), ecumenical dialogue across the global divides is conspicuous in its absence.

This volume and this series as a whole are born from a recognition that such challenges must be addressed but that this is much easier said than done. This is expressed in the subtitle of the series, namely "Telling the Story amid the 'Anthropocene.'" There are multiple stories, but such stories of who God is and what God is doing must be somehow related to each other, or else we may end up speaking of different gods and different planets. Moreover, the danger is that one may separate God's work of creation from other aspects of God's work, such as salvation and consummation. The ecumenical challenge is then to speak not *only* but *also* of "the story" in the singular. The challenge is exacerbated by the so-called Anthropocene and the contestations over naming it as such. Again: Who is the implied Anthropos? Whose story is it? Can Christians tell this story together, to each other?

■ Conclusion: The Question is ... What is the Question?

It should be clear from the above that creation theology, also in the context of Christian ecotheology, tends to go in widely diverging directions. Here is one simplified way of capturing such directions.

Some focus on the question of *whether* the world was indeed created, typically in conversation with philosophy and astrophysics. Some focus on the question of *how* the world (or life on Earth and humans in particular) was created, typically in conversation with the natural sciences, including evolutionary biology. Some argue that the focus should be, instead, on *who* created it and therefore engage in doxology—or in dialogue—with other religious traditions. Some follow the more traditional approach to focus on *what* God created and then engage in theological reflection on stars, plants, animals, evolution, being human (the image of God), racial and ethnic diversity, the visible and the invisible, time and space, the angels, or even astrobiology and the search for extraterrestrial intelligence—with all the dangers that this may entail. One may also speculate on *why* God created the world and reflect on the purpose of creation. Yet others explore the question of *what it may mean to "create"* and reflect on the differences between being and becoming, order and chaos, generating and birthing, Word and Spirit, productivity and letting be, God's role and the earth's own creativity, blessing and being blessed, working and resting in God's grace.[36] Finally, one may also focus on *the meaning of being created*, celebrating a sense of inhabitation and cohabitation, of being embodied and of symbiosis with other bodies in God's presence. Such a focus on embodied inhabitation may then also provide a point of entry for a critique of violations of the integrity of creation.

The full diversity of approaches and directions will come into play in this volume. In order to allow some coherence to the volume and to engage in further conversation, each contributor has been asked to focus on the same question. It may help to state that again: "What difference does it make to the story of cosmic, planetary, human, and cultural evolution to re-describe this as the creative work of God's love? Inversely, what difference does it make to the story of God's love to describe it in evolutionary and geographic terms?"

There is no need in this editorial introduction to provide an overview of the various essays included in this volume. Each contributor will speak for themselves from within a particular confessional tradition, theological school, and geographical context. The contributions are therefore simply placed in alphabetical order. The volume will conclude with a conversation between all the contributors on current paths and emerging horizons in creation theology, with specific reference to Christian ecotheology. The purpose is not to close off the conversation but to capture something of the current state of the debate and to stimulate further reflection through a set of contributions with an optimized diversity.

■ Bibliography

Abrecht, Paul, ed. *Faith, Science and the Future*. Geneva: WCC, 1978.

Abrecht, Paul, ed. *Faith, Science and the Future in an Unjust World, Volume 2: Reports and Recommendations*. Geneva: WCC, 1980.

Altner, Günter. *Schöpfung am Abgrund: Die Theologie vor der Umweltfrage*. Neukirchen-Vluyn: Neukirchener Verlag, 1974.

———. *Zwischen Natur und Menschengeschichte: Anthropologische, biologische, ethische perspektiven für eine neue Schöpfungstheologie*. München: Kaiser Verlag, 1975.

Baker-Fletcher, Karen. *Sisters of Dust, Sisters of Spirit: Womanist Wordings on God and Creation*. Minneapolis: Fortress, 1998.

---

36. For juxtaposing these questions, see Conradie, "What on Earth did God Create?"

Bhagwan, James, et al., eds. *From the Deep: Pasifiki Voices for a New Story*. Suva: Pacific Theological College, 2020.

Black, John. *The Dominion of Man: The Search for Ecological Responsibility*. Edinburgh: Edinburgh University Press, 1970.

Boff, Leonardo. *Cry of the Earth, Cry of the Poor*, Maryknoll: Orbis, 1997.

Boff, Leonardo, and Virgilio Elizondo. *Ecology and Liberation: A New Paradigm*. Maryknoll: Orbis, 1995.

Boff, Leonardo, and Virgilio Elizondo, eds. *Ecology and Poverty. Cry of the Earth, Cry of the Poor. Concilium*. London: SCM, 1995.

Bouma-Prediger, Steven. *For the Beauty of the Earth: A Christian Vision for Creation Care*. 2nd ed. Grand Rapids: Baker Academic, 2007.

Castillo, Daniel P. *An Ecological Theology of Liberation: Salvation and Political Ecology*. Maryknoll: Orbis, 2019.

Clough, David. *On Animals Volume I: Systematic Theology*. London: T&T Clark, 2012.

———. *On Animals Volume II: Theological Ethics*. London: T&T Clark, 2019.

Cobb, John B., Jr. *Is It Too Late? A Theology of Ecology*. Beverly Hills: Bruce, 1971.

Cone, James H. "Whose Earth Is It Anyway?" In *Earth Habitat: Eco-injustices and the Church's Response*, edited by Dieter T. Hessel and Larry Rasmussen, 23–32. Minneapolis: Fortress, 2001.

Conradie, Ernst M. *Christianity and a Critique of Consumerism: A Survey of Six Points of Entry*. Wellington: Bible Media, 2009.

———. *Christianity and Earthkeeping: In Search of an Inspiring Vision*. Stellenbosch: SUN, 2011.

———. "Doing Creation Theology in the South African Context." *Journal of Systematic Theology* 1:6 (2022), 1–37.

———. *The Earth in God's Economy: Creation, Salvation and Consummation in Ecological Perspective*. Berlin: LIT Verlag, 2015.

———. "On Social Evil and Natural Evil: In Conversation with Christopher Southgate." *Zygon: Journal of Religion and Science* 53:3 (2018), 752–65. https://doi.org/10.1111/zygo.12427

———. *Saving the Earth? The Legacy of Reformed Views on "Re-Creation"*. Berlin: LIT Verlag, 2013.

———. "What on Earth did God Create? Overtures to an Ecumenical Theology of Creation." *The Ecumenical Review* 66:4 (2014), 433–53.

Conradie, Ernst M., ed. *Creation and Salvation, Volume 1: A Mosaic of Essays on Selected Classic Christian Theologians*. Berlin: LIT, 2011.

———. *Creation and Salvation, Volume 2: A Companion on Recent Theological Movements*. Berlin: LIT, 2012.

———. "Ecumenical and Ecological Perspectives on the 'God of Life'." *The Ecumenical Review* 65:1 (2013), 1–2, 3–165.

Conradie, Ernst M., et al., eds. *Christian Faith and the Earth: Current Paths and Emerging Horizons in Ecotheology*. London: T&T Clark, 2014.

Conradie, Ernst M., Sipho Mtetwa, and Andrew Warmback eds. *The Land is Crying for Justice: A Discussion Document on Christianity and Environmental Justice in South Africa*. Stellenbosch: Ecumenical Foundation of Southern Africa, 2002.

Daly, Mary. *Gyn/Ecology: The Metaethics of Radical Feminism*. Boston: Beacon, 1978.

Daneel, Marthinus L. *African Earthkeepers: Wholistic Interfaith Mission*. Maryknoll: Orbis, 2001.

Deane-Drummond, Celia E. *Eco-Theology*. London: Darton, Longman & Todd, 2008.

Edwards, Denis. *Ecology at the Heart of Faith: The Change of Heart that Leads to a New Way on Earth*. Maryknoll: Orbis, 2006.

Evangelische Kirche in Deutschland, Advisory Commission on Sustainable Development. *Livestock and Fellow Creatures! Animal Welfare, Sustainability and the Ethics of Nutrition from a Protestant Perspective.* Hannover: EKD, 2019.

Fox, Matthew. *Original Blessing.* New Mexico: Bear & Co., 1983.

Gebara, Ivone. *Longing for Running Water: Ecofeminism and Liberation.* Translated by David Molineaux. Minneapolis: Fortress, 1999.

Gitau, Samson K. *The Environmental Crisis: A Challenge for African Christians.* Nairobi: Acton, 2000.

Gregorios, Paulus. *The Human Presence: Ecological Spirituality and the Age of the Spirit.* New York: Amity, 1987.

Grey, Mary. *Redeeming the Dream: Feminism, Redemption and Christian Tradition.* London: SPCK, 1989.

Gunton, Colin. *The Triune Creator: A Historical and Systematic Study.* Grand Rapids: Eerdmans, 1998.

Gustafson, James M. *A Sense of the Divine: The Natural Environment from a Theocentric Perspective.* Cleveland: Pilgrim, 1994.

Halkes, Catharina J.M. *Christian Feminism and the Renewal of the Earth.* London: SPCK, 1991.

Hall, Douglas John. *The Steward: A Biblical Model Come of Age.* Grand Rapids: Eerdmans, 1990.

Hallman, David G., ed. *Ecotheology: Voices from South and North.* Geneva: WCC, 1994.

Harris, Melanie. *EcoWomanism.* Maryknoll: Orbis, 2017.

Haught, John F. *The Promise of Nature: Ecology and Cosmic Purpose.* Mahwah: Paulist, 1993.

Hefner, Philip. *The Human Factor: Evolution, Culture and Religion.* Minneapolis: Fortress, 1993.

Hendry, George S. *Theology of Nature.* Philadelphia: Westminster, 1980.

Jantzen, Grace. *God's World, God's Body.* Philadelphia: Westminster, 1984.

Jennings, Willie James. "Reframing the World: Toward an Actual Christian Doctrine of Creation." *International Journal of Systematic Theology* 21:4 (2019), 388–407. https://doi.org/10.1111/ijst.12385

Johnson, Elizabeth A. *Women, Earth and Creator Spirit.* New York: Paulist, 1993.

Kanyoro, Musimbi R., and Nyambura J. Njoroge, eds. *Groaning in Faith: African Women in the Household of God.* Nairobi: Acton, 1996.

Kaoma, Kapya J. *God's Family, God's Earth: Christian Ecological Ethics of Ubuntu.* Blantyre: Kachere Series, 2014.

Kim, Grace Ji-Sun, and Hilda P. Koster, eds. *Planetary Solidarity: Global Women's Voices on Christian Doctrine and Climate Justice.* Minneapolis: Fortress, 2017.

Liedke, Gerard. *Im Bauch des Fisches: Ökologische Theologie.* Stuttgart: Kreuz Verlag, 1979.

Link, Christian. *Schöpfung Band I: Schöpfungstheologie in reformatorischer Tradition.* Gütersloh: Gütersloher Verlagshaus, 1991a.

———. *Schöpfung Band II: Schöpfungstheologie angesichts der Herausforderung des 20. Jahrhunderts.* Gütersloher Verlagshaus, 1991b.

McFague, Sallie. *The Body of God: An Ecological Theology.* London: SCM, 1993.

Mendoza, S. Lily, and George Zachariah, eds. *Decolonizing Ecotheology: Indigenous and Subaltern Challenges.* Eugene: Pickwick, 2021.

Moltmann, Jürgen. *God in Creation: An Ecological Doctrine of Creation.* London: SCM, 1985.

Mosala, Itumeleng, and Buti Thlagale. *The Unquestionable Right to be Free: Essays in Black Theology.* Braamfontein: Institute for Contextual Theology, 1986.

Mugambi, Jesse N.K. *God, Humanity and Nature in Relation to Justice and Peace*. Geneva: WCC, 1987.

Mugambi, Jesse N. K., and Mika Vähäkangas, eds. *Christian Theology and Environmental Responsibility*. Nairobi: Acton, 2001.

Nash, James A. *Loving Nature: Ecological Integrity and Christian Responsibility*. Nashville: Abingdon, 1991.

Ruether, Rosemary Radford. *Gaia and God: An Ecofeminist Theology of Earth Healing*. New York: Harper & Collins, 1992.

———. *New Woman/New Earth: Sexist Ideologies and Human Liberation*. New York: Seabury, 1975.

Ruether, Rosemary Radford, ed. *Women Healing Earth: Third World Women on Ecology, Feminism and Religion*. Maryknoll: Orbis, 1997.

Santmire, H. Paul. *The Travail of Nature: The Ambigous Ecological Promise*. Philadelphia: Fortress, 1985.

Sideris, Lisa H. *Consecrating Science: Wonder, Knowledge and the Natural World*. Oakland: University of California Press, 2017.

Sittler, Joseph. *Essays on Nature and Grace*. Philadelphia: Fortress, 1972.

Sölle, Dorothee. *On Earth as in Heaven: A Liberation Spirituality of Sharing*. Louisville: Westminster John Knox, 1993.

Sölle, Dorothee, and Cloyes, Shirley. *To Work and to Love: A Theology of Creation*. Philadelphia: Fortress, 1984.

Southgate, Christopher B. *The Groaning of Creation: God, Evolution, and the Problem of Evil*. Louisville: Westminster John Knox, 2008.

———. *Theology in a Suffering World: Glory and Longing*. Cambridge: Cambridge University Press, 2018.

Swimme, Brian, and Mary Evelyn Tucker. *Journey of the Universe*. New Haven: Yale University Press, 2011.

Swimme, Brian, and Thomas Berry. *The Universe Story: From the Primordial Flashing Forth to the Ecozoic Era*. New York: Penguin, 1992.

Tofaeono, Ama'amalele. *Eco-Theology: Aiga – The Household of Life: A Perspective from Living Myths and Traditions of Samoa*. Erlangen: Erlanger Verlag für Mission und Ökumene, 2000.

Townes, Emilie M., et al., eds. *Walking Through the Valley: Womanist Explorations in the Spirit of Katie Geneva Cannon*. Louisville: Westminster John Knox, 2022.

Vaai, Upolu Lumā. "We Are Earth: reDirtifying Creation Theology." In *reSTORYing the Pasifika Household*, edited by Upolu Lumā Vaai and Aisake Casimira, 63–84. Suva: Pacific Theological College Press, 2023.

Welker, Michael. *Creation and Reality*. Minneapolis: Fortress, 1999.

Westhelle, Vítor. "Creation Motifs in the Search for a Vital Space: A Latin American Perspective." In *Lift Every Voice: Constructing Christian Theologies from the Underside*, edited by Susan Brooks Thistlethwaite and Mary Engel Potter, 146–58. Maryknoll: Orbis, 1998.

Zachariah, George. "Whose Oikos Is It Anyway? Towards a Poromboke Eco-theology of 'Commoning'." In *Decolonizing Ecotheology: Indigenous and Subaltern Challenges*, edited by S. Lily Mendoza, and George Zachariah, 205–22. Eugene: Wipf & Stock, 2021.

Treaty as a Shared Narrative: Indigenous Treaty as Canada's Creation Story

Ray Aldred[1]

■ Introduction

This essay flows out of the writing of my thesis, albeit an idea I only mentioned briefly, namely, that the historic treaty was an attempt by Indigenous people to narrate the newcomers into the story of the land or into Indigenous communal identity. The main premise of my thesis, entitled "An Alternative Starting Place for an Indigenous Theology," was that if you began with Indigenous communal narrative identity, this would invigorate theology in ways that have been stunted in modern evangelical theology.

Western theology, or at least popular Western theologies, began with the alienated individual of modern thought. Indigenous scholar Vine Deloria Jr. noted that Western thinking focuses on the individual: "From John

---

1. Ray Aldred is status Cree from Swan River Band, Treaty 8, in Northern Alberta, Canada. He is the director of the Indigenous Studies Program at the Vancouver School of Theology. He is registered as a coresearcher at the University of the Western Cape for the project on "An Earthed Faith: Telling the Story amid the 'Anthropocene.'"

> **How to cite:** Aldred, R 2024, "Treaty as a Shared Narrative: Indigenous Treaty as Canada's Creation Story", in EM Conradie & WJ Jennings (eds.), *The Place of Story and the Story of Place,* in An Earthed Faith: Telling the Story amid the "Anthropocene", vol. 3, AOSIS Books, Cape Town, pp. 25–40. https://doi.org/10.4102/aosis.2024.BK355.02

Locke to John Rawls, the important decisions are to be made by individuals possessing neither father nor mother, village nor tribe, age nor gender."[2]

Western theology, then, seems to begin with a similar starting point of the alienated individuals found in Genesis 3,[3] with the effects of the curse impacting every relationship: relationships with human beings, relationships with the Creator and spiritual beings, relationships with the land itself. Sin and the need for salvation were caricatured in some evangelical theologies to focus almost completely on the individual salvation of human beings. Starting in Genesis 3 meant that human beings were alienated from everyone and everything. This meant that the world was a dangerous, demonic place that one needed to wage war against. Perhaps this was why John West, an early missionary to Western Canada, stated that the heathen needed to be converted and the heath needed to be cultivated.[4] The wild people and wild land needed to be tamed. Early on, much of the North American gospel became marked by a message of prosperity. All one had to do was receive Jesus, and this would then usher in a new period where people had lots of friends, complete healing, and lots of money. Your churches would grow, your influence would grow, and through this steady advance of individual salvation, you would win the day and fly away to God's celestial shore, someplace out there. The preoccupation, it seems to me, was with the solitary, autonomous individual.

I thought that if you began with Indigenous communal identity, you might get to a different place. Indigenous identity tended to be communal because we believe that we are related to the earth. The Cree cosmology begins with, "It is a good world." An Indigenous theology begins with harmony, not with alienation. Not a harmony that is apathetic and devoid of tension but an active harmony that must be stewarded. It is a harmony that begins with a good world that the Creator has placed us upon. The boundaries of our Indigenous world are maintained by our stories. The stories we tell help us see the unseen and understand our place or the space we occupy, as well as giving us the laws that guide us to understand how we are to live out this harmony that has been given to us. As I was working on my thesis, I was struck by how the historic Indigenous treaty-making was an outflow of this Indigenous communal identity.[5] A communal identity that did not devolve into tribalism but was open to the newcomer and, contrary to our current cultural milieu that seems intent on trying to

---

2. See Deloria, "Philosophy and the Tribal Peoples," 10.

3. I understand that the "fall" is better understood as covering Genesis 3 to 11. Sin spreads and society seems unable to stem the tide.

4. See Austin and Scott, *Canadian Missionaries*, 22.

5. See Miller, "Compact, Contract, Covenant."

block out all those who are different from us, seemed to make them enemies and establish laws to keep them out. Indigenous treaty-making was a process that seemed to flow from the harmony that flowed from creation.

I have also been thinking about the tasks that the church could engage in that would contribute to our journey toward reconciliation. As I have already stated, the preoccupation with individual salvation by the neo-Christendom church has meant that popular theology in much of North America has little to do with reconciliation between people and people groups but is limited to individuals and God. Having said this, there are some good examples of churches that have understood that salvation is all about Christ reconciling all things, Creator and creation in perfect harmony. Yet many within the church spend a significant amount of time arguing about which church is the greatest, or as we in the dominant church have spent much of our time, pointing out our distinctives, which usually translate into denominational distinctives. These are usually answers to questions that no one is asking.[6] We quickly forget that the canons of the church are not there to bind but to help discern how we live out our lives in relationship with all our relatives.[7] Christians, however, have tended to interpret the canon as law to suppress insurrection against church authority. Instead, the question that people have been asking of the church, during this time when we are finding the unmarked graves of residential school survivors and marginalized people revealing the multigenerational trauma in the middle of which the church is situated, is this: "Where is God in all of this?"

How do we heal in Canada? In the current situation in Canada, the Indigenous church asks, "How do we transform ourselves through healing?" The non-Indigenous church asks, "What should we do?" Newcomers rarely reflect on who they should be and focus on transformation through healing. I used to try to get people to shift to think about their identity in a different way, but perhaps this is just part of the newcomer language game. Therefore, whenever I speak in non-Indigenous settings and the time for questions comes, people tend to ask: "What should I do?" So, here are three categories of things people can do.

■ Heal the Land

The revealing of unmarked graves could be heard as a prophetic call. Howard Jolly, the director of the First Nations Alliance Churches of Canada,

---

6. I was riffing off of Andrew Walls's idea that Western missionaries tended to teach theology by giving answers to theological questions no one was asking. For example, proofs for God's existence are pointless to Indigenous people, who do not question that there is a creator.

7. I owe this insight to Mark MacDonald, former Archbishop for the Indigenous Anglican Church in Canada.

wrote a song about the unmarked graves, connecting their finding to the words of God to Cain, namely that your brother's blood cries to me from the ground. The wound upon the First Peoples is a wound upon the land, and a wound upon the land is a wound upon human beings.

We need to be working to heal the land. The challenge that Indigenous thought and vision poses to dominant Canadian society and the church is to explain what this means. When I talk about healing the land, people immediately jump to the category of ecology or Earth care. This is somewhat helpful, or it may be just a diversion or a way to salve the legitimate shame that continues to needle the soul of Canada as they fail to honor the treaties and continue to attack the relationship that human beings, Indigenous people, have with the land. There needs to be a wholistic approach to healing the land.

■ Develop or Raise Emotional Intelligence and Skill

The multigenerational trauma that Indigenous people have been through has not been part of the mainstream theological conversation of the church, although it has been part of the conversation in the majority church of the Two-Thirds World, who have been subjected to colonialism and globalization or Westernization. Popular North American theology has continued to proliferate a theology that promises prosperity, using violence to advance these ends.

In Canada, at least until the finding of unmarked graves near former residential schools, the church engaged in a form of gaslighting, blaming the economically and socially marginalized for their own problems, blaming them for not embracing the prosperity gospel that would solve all their problems.

The other form of denial engaged in by the church is to make half-healed people (a term from Renée Altson's book)[8] pretend that everything is okay before they can participate in ministry in the church. Even though we follow a savior who is a man of sorrows and acquainted with grief, the church in North America, because of its spirituality of numbness, lacks the emotional intelligence to provide a theology that sees the damaged individual as having anything to offer. Dr. Martin Brokenleg has talked to us Indigenous people about the need to rebuild our emotional intelligence so that we can continue to speak from the heart and help our children and grandchildren develop the emotional resources to live in our world.[9] The Western church,

---

8. See Altson, *Stumbling Toward Faith*.

9. See Martin Brokenleg, "The Spirituality of Self-Determination."

however, tends to rely on sin management techniques[10] and forms of escapism through consumer-driven ministries.[11] We treat damaged people as problems to be solved, with the result that most Canadians have been paternalistic at best in their thinking and treatment of Indigenous people. We need to develop emotional intelligence.

■ Cast a Vision of Unity That is Neither Tribalism nor a New Imperialism

Paul Ricoeur speaks of embracing the intended unity and harmony that the Creator has planned for all things.[12] He takes aim at the nation-state as a hindrance to this unity. From that essay, he goes on to say that some nations have assumed that they were the kingdom of God come to Earth. As a result, as Lesslie Newbigin alludes to in *Foolishness to the Greeks: The Gospel and Western Culture*,[13] they brought hell instead. We need to embrace that it is God's intention that we move toward one another and live in harmony with the Creator and all creation. The problem with nation-states (and, I think, denominations, too) is that they presume that their particular expression of Christian faith is the one that God intended for everyone. I am not against denominations per se, but in their institutional form, they tend toward being a power and principality that uses coercion as a way of maintaining control. The Truth and Reconciliation Commission of Canada (TRCA) raises this issue in Action Item 60 in calling for churches to educate clergy to stop engaging in religious violence by dividing communities along religious lines.[14] We need to cast a vision that affirms the personhood of all human beings.

The call to pursue reconciliation comes primarily from Indigenous people because they see the strain of trauma on their communities. Perhaps where there are communities of people who live insular lives and have achieved or perfected the ability to remove tension from their daily existence, the goal of the modern good life,[15] reconciliation and unity seem too much to pursue, but the elders continue to put forward the need to pursue and affirm the treaty relationship. This is evident in the

---

10. See Willard, Harney, and Harney, *The Divine Conspiracy*.

11. See Colwell, *Promise and Presence*, 12–13.

12. See Ricoeur, *Political and Social Essays*, 141–48.

13. See Newbigin, *Foolishness to the Greeks*, 116–17.

14. See Truth and Reconciliation Commission of Canada, *Calls to Action*, §8.

15. See Ellul, *The Technological Society*.

TRCA's final report[16] and is affirmed by my experience in listening to the discussions.

Having said this, the church must give up its need to be in control of the process of reconciliation. Again, because the only power available to most head offices of institutions is the power of coercion, they tend to use that power to enforce unity. This is a form of imperialism or neocolonialism, and, in the churches' case, shaming Indigenous efforts at self-determination by telling Indigenous people not to undermine unity within the denomination. This kind of rhetoric is identified by Timothy Schouls.[17] Canadian politicians have a history of claiming that Indigenous communal identity is an enemy of Canadian individual identity.

There is a need to work toward unity that does not devolve into tribalism, as defined by Amartya Sen or Tsvetan Todorov.[18] The latter does a good job of pointing out that some forms of conservativism are enemies of democracy[19] and unity because they enforce a particular form of identity upon all members of the human race, one that their own conservatism defines and dictates and that, interestingly enough, is an identity that mirrors themselves. Karl Barth would warn us of the dangers of declaring ourselves righteous and everyone else unrighteous.[20]

We have, I believe, in Canada an opportunity to build a unity within this country upon two resources, both rooted in the identity, innate spirituality, or sense of sacredness of Indigenous people, namely the historic treaty process and creation. The focus of this essay will be on spirituality and treaty as a shared narrative and on Canada's creation story.

■ Identity, Spirituality, and Law from Creation Stories

Identity

If you listen to Indigenous creation stories, they speak of their relationship to the earth. They speak of how they came to be upon the land or upon the sea. When I use the term 'land,' I am using it as shorthand for creation. Creation is the context for life; there is no life apart from creation. This simple truth permeates Indigenous understanding. Vine Deloria Jr. and

---

16. See Truth and Reconciliation Commission of Canada, *Calls to Action*, §45.

17. See Schouls, *Shifting Boundaries*, 37.

18. See Sen, *Identity and Violence*, 9–10.

19. See Todorov and Brown, *Inner Enemies of Democracy*, 9–11.

20. See Barth, *Church Dogmatics IV*, 60–62.

other Indigenous scholars note that Indigenous philosophy is based primarily upon our interaction with all our relatives.[21] The land is full of life, and in Cree, the word for spirit and life is the same word because all of life is full of spirit. I am not debating, at this point, whether this means that trees have souls, but I understand that the trees communicate with one another. On a recent tour of the Botanical Gardens at the University of British Colombia, it was explained that the tree whose European name is the Douglas Fir will communicate to the other trees that fire is close, and the rest of the Douglas Firs will begin to develop a natural fire retardant. One elder said to me that we understand that even a rock is full of life, and if you look at it under a microscope, you see all the life that is living on the rock. Our identity flows from our creation stories.

The creation stories are handed down to us from our elders. The elders have local, personal knowledge of the land and teach us about the land.[22] This local knowledge of the land includes their experiences of residential schools and of the oral history of the specific treaty negotiations, all of which prove quite accurate at points.

The elders also relate to us the mythical stories that take in all time and all space. This does not mean, however, that these stories are unreal, so we do not call them myths, but we call them legends. Together, these stories taught by our elders give us a shared communal memory that leads to a shared ontology.[23] As I have already stated, this is contrary to a Western worldview, particularly an academy that leans toward viewing religion through the lens of the phenomenology of an autonomous individual and relegates all spirituality to the margin of reality or to the realm of the unreal and invisible. Religion and knowledge, then, are moved to the inner world of abstract knowledge. The problem with the academy and its world of abstract knowledge has been its inability to write in a way such that most folks in society could see its relevance. This is pointed out by J.R. Miller in writing about the reconciliation process,[24] namely that the academics (theologians and biblical studies scholars) do not write in a manner that is accessible to people. As a result, we have not been as helpful in the reconciliation process as we might have been.

The beauty of creation stories is that as we understand how we are related to the earth, "we can feel the earth welcome us home."[25] I had the privilege of listening to the Stoney Nakoda creation story of how they came

---

21. See Deloria, "Philosophy and the Tribal Peoples," 10.

22. See Kovach, *Indigenous Methodologies*, 143-44.

23. See Wilson, *Research is Ceremony*, 57.

24. See Miller, *Residential Schools and Reconciliation*, 266.

25. See McLeod, *Cree Narrative Memory*.

out of the mountain in what is now called Banff. I have heard the Lakota creation story of how they came out of a cave and came to be on their territory. And I have heard different versions of how we Cree came to be where we are on what we call Mother Earth or Turtle Island. What is common is that in each instance, we are in the middle of a good world.

I believe that all people who live in this place need to experience or can experience the land welcoming them home. A Cree elder from Northern Alberta is quoted in a book of stories as saying that the reason why Euro-Canadians have so many problems is that they have lost touch with their connection with the land.[26] This must be why they want to name the land after themselves or after some place where they used to live.

Creation stories form our identity. Stanley Grenz wrote that Scripture becomes a shared story for developing an ecclesial identity.[27] Canada, however, needs a new creation story because the imaginary story of Canada as a benevolent Christian power that settled and improved the land and helped the Indigenous people dig themselves out of lives that were "brutish, short and bereft of hope" does not work anymore. There have been attempts to develop Canadian shared narratives. There was one put forward by the former prime minister of Canada, Stephen Harper, and others, namely that Canada was a country of explorers.[28] Harper's government worked hard to shift Indigenous issues from the historic relationship of nation to nation, to Indigenous issues being a subset of Canadian issues. Harper, in my opinion, was always underplaying the role of Indigenous people in Canada's identity. For him, the findings of British explorers showed that Canada has a history of showing sovereignty over the Arctic.[29] It is interesting that in his history of Canada, he does not mention the role of Indigenous people living in the Arctic, which shows their identity is older than Canada as a nation-state and older than the tragic Franklin expedition.

There are other explorer myths. My daughter, Catherine, found an account of David Thompson, who was searching for the ocean and was impatient with his Indigenous guides because he believed that they should have arrived at the ocean at a certain point. The Indigenous guides responded by telling him, "Well, why don't you lead, since you know everything."

---

26. See Meili, *Those Who Know: Profiles of Alberta's Native Elders.*

27. See Grenz, *Theology for the Community of God,* 500–501.

28. See Harper, "The Call of the North,"

29. See John Geiger, "Prime Minister Stephen Harper on the Franklin Find," Canadian Geographic Enterprises, November 3, 2014, https://canadiangeographic.ca/articles/prime-minister-stephen-harper-on-the-franklin-find/.

Returning to the Franklin expedition, Indigenous oral tradition knew where the ships had sunk, but none of the Europeans were asking.

We, therefore, also need a new national narrative.

Spirituality

With all these terms—identity, spirituality, and law—even though I am using them as nouns, I understand them as containing a dynamic process that is always moving and changing. Spirituality is dynamic among us Cree. Perhaps this can be seen in our understanding of the great mystery, that we use the word *gitchi manitou*, which is sometimes translated as 'Creator,' 'mystery' or 'big something'; it is an understanding that our perspective is small, and we are always seeking understanding to live in a good way upon a good world.

The Cree cosmology begins with an acknowledgment that this is a good world. Harmony is built into creation, and we human beings are the least wise among the creatures and beings that inhabit the earth, but we are called to be stewards of our land. We are the least wise because while the creatures all understand by instinct how they are to live, we humans must be taught and learn to harness our free will, for we are creatures that cross over various boundaries at different times.

Spirituality, then, is not like the Western concept of religion, although some have taken this route as a way to at least gain an equal hearing for Indigenous spiritualities. Religion is then relegated to the realm of an interior dialogue that individuals have with themselves and has limited value as a source of a common story for everyone.

Spirituality is about living life in good relationships. I have borrowed four categories of relationship from Francis Schaeffer,[30] namely relations with spiritual beings, relations with the land or creation, relations with our group and other groups, and relations with oneself. Spirituality is about living out the harmony that is all around us in creation. Creation itself teaches us how to live in a good way, but we need the elders to help us develop a vision for living on the land.

The idea of crying for a vision among the northern plains people captures this idea. When a person comes of age or they are faced with challenging events, they go out on the land and cry for a vision. They are searching for a vision of their authentic self.[31] They abstain from food and water and stay in the same place for four days. On the third day, the demons come, and

---

30. See Schaeffer, *Pollution and the Death of Man*, 6.

31. See Elk Black and Brown, *The Sacred Pipe*, 44-66.

one must battle through this, and on the fourth day, one receives a vision that will define one's life. One returns and the elders help one to interpret what this vision means.

This "vision quest" and all our ceremonies are understood to flow out of our creation story, which flows from our communal memory and communal stories that tell of our relationship with the land. Our creation stories tell us who we are and form the basis of our spirituality.

In all of these cases, people embrace a communal identity and spirituality that is tied to the land. The land is sacred because, at any point, the Creator could do something powerful. The spirituality then flows out of the vision upon the land. It is a communal spirituality with the presence of the elders, who help to locate this vision within the communal memory and story of the people upon the land.

Law

Not every Indigenous group performs the vision quest ceremony, but Indigenous law also flows from creation (or the land). According to John Borrows, law has a metaphysical element in that it is related to our creation stories.[32]

Western colonization was premised upon the idea that the "native" or "primitive" people of the New World did not possess laws. Indigenous people were thought, therefore, to be less than human, but they could be raised to the level of being citizens of the new nation-state. The Western concept of law was therefore imposed upon them. In Canada, for example, the "Indian Act" was used by the Canadian government to make legal what would be illegal or at least immoral. The *Indian Act* was used by the Canadian government to separate children from their parents so that the children could be placed in residential schools to be assimilated into Euro-Canadian society. This concept of law, which I call legality, was also used as justification to take and remain in control of unceded land by declaring both the doctrine of discovery in the case of the Americas and *terra nullius* in Australia and British Colombia. As a result, Western law has become completely dominated by litigation and the arguments for individual property rights and ignores the communal identity of Indigenous nations and people.

Indigenous law, on the other hand, flows from creation. It helps us to understand how we are related to all things and how our freedom is a gift for the community. This can be seen in many of the ceremonies. The seeking

---

32. See Borrows, *Recovering Canada*, 5–15.

of a vision is so that you can better fit within the community or take your place for the community. Anthropologist Richard Preston, in writing about Cree narratives, observed that the Cree have a decentralized government that does not lead to anarchy.[33] They control their world, he says, by mental force. Each person grows up with the understanding of how each choice they make impacts every other relationship in their world. This results in a society that abhors the use of violence against another human being, where a chief would diffuse tense situations by being the most gracious, generous, well-mannered individual in the room.

True spirituality in an individual is identified by how they live in relationship with all things. It is not primarily based upon the words that people say but on the lives that people live. We can see this in the example of how an apology is supposed to work. Eva Mackey states that apology is supposed to lead to a conversation, not to shutting down conversation.[34] An apology, according to the Cree elders, means that you make amends. Regardless of the words used in the apology from former prime minister Stephen Harper, there was not an apology, according to the elders of one Northern community, because the actions of those who apologized had not changed and they had not made the necessary amends to their lives.[35]

Indigenous identity, spirituality and law flow out of the story of their life upon the land and then lead naturally to the idea of a treaty as a way to maintain respectful relationships with other First Nations and with the newcomers who came to be upon the land. Indigenous people have a history of making treaties as a way to share place that stands in sharp contrast to the propensity of modern nations to try and enforce boundaries by imposing more security.

■ A Treaty Is an Attempt by First Nations at a Shared Narrative

When the newcomers came, it ushered in a time of significant change. First Nations leaders, as well as the Métis and later the Inuit, would all see the necessity of forming relationships with the newcomers.

Remember that spirituality is built upon creation stories and that from these creation stories we understand that the land is our relative. Understanding that, in the fall, you pray for good hunting. You make it through the winter, and you give thanks for good hunting. Early missionaries

---

33. See Preston, *Cree Narrative*, 78.

34. See Mackey, "The Apologizers' Apology," 48–49.

35. Personal conversation with Treaty 8 Grand Chief Arthur Noskey.

write, however, that the Indigenous people did not hunt the animals but believed that the animals gave themselves to them. For this, the animal was thanked, and this thankfulness was to safeguard respect and honor.

The land and animals are relatives, but we do not own the land in the way the Europeans thought they could own the land. It is not that Indigenous people did not have possessions, but Arthur Noskey, Grand Chief of Treaty 8, told me recently that you can only possess what you can carry on your back. That is why we say we do not carry the land, but the land owns us. We do not own the land; the land owns us.

Our spirituality and creation story demand that we "tell" the newcomers into the story. I use the word "tell" because "write" refers to a textual culture, but Indigenous culture remains largely shaped by orality. Oral tradition means that you do not write someone into the story; you tell them into the story. Our treaties are not primarily the words on a page but the stories and lives lived out on the land together. As for the coming of the newcomers, numerous prophets in Indigenous communities had seen the Europeans coming. So when they came, we made treaty.[36]

J.R. Miller, writing about treaty, proposes that the Indigenous treaty process slowly developed and shifted.[37] Early on, the relationship between the newcomers and Indigenous people centered around the economic and political desires that the newcomers had and for which they needed the assistance of Indigenous people. Indigenous people were participants in these treaties because they helped to further Indigenous aspirations.[38] The issue of land, however, and Indigenous title to the land was addressed by the Royal Proclamation of 1763. The Royal Proclamation stated that Indigenous people had the land and that no individual or organization, except for the British Crown, had the authority to negotiate issues of land with Indigenous people.

Early in the settlement of newcomers to the land now called Canada, simple compacts about the use of land were entered into by newcomers and Indigenous people. Later, newcomers who wanted access to land negotiated contracts with Indigenous people around the Great Lakes.[39] This second phase also introduced the idea of annuities, as the British government realized that it did not have the resources to buy the land wanted; therefore, instead of making lump-sum payments for the land, the British counted on tax dollars raised from settlers to pay the annuity to the

---

36. To "make treaty" is a common expression in the Canadian context (editor).

37. See Miller, "Compact, Contract, Covenant."

38. See Miller, "Compact, Contract, Covenant," 4.

39. See Miller, "Compact, Contract, Covenant," 15.

First Nations for being able to use the land.[40] Treaty-making continued to shift because the situation had shifted. Indigenous people on the Prairies entered into agreements resembling covenants between the newcomers, Indigenous people, and the Creator. The involvement of the Creator was what made the treaty like a covenant. We promised to live like relatives, each group participating in one another's ceremonies, and the Creator would hold us to it. The treaties are renewed every year and are in effect as long as rivers run.

The historic Indigenous treaty therefore affirms at least four basic principles. The first three I heard from Grand Chief Stan Beardy of the Nishnawbe-Aski Nation, and I added a fourth that I heard from former Indigenous Archbishop Mark MacDonald:

- The privilege of a peaceful existence.
- The privilege of access to the land.
- The privilege of being fed from the bounty of the land.
- The privilege of being who the Creator made us to be (added by Mark MacDonald).

■ Spirituality is Built upon the Understanding of a Good World

Remembering that the goal of spirituality is built upon the foundation of a good world, Indigenous people made a treaty which eventually took the form of a covenant. It took a covenantal form because the participants involved were the First Nations, the newcomers, and the Creator. Some of the signatories from the Lakota and Dakota understood that in the treaty process, we were becoming relatives.

Indigenous spirituality, which was communal and based upon shared stories or communal narratives that included land, seemed to be able to embrace migration from the very beginning. We were travelers on a journey. We would live upon the earth, but after we were gone, the earth would remain. So when the newcomers came, Indigenous people embraced these newcomers by making relatives, which they understood would change their identity but that it would help them to be who they were created to be, humans living out their lives, properly related to all creation.

Black Elk, a Catholic catechist and Lakota holy man, explained his understanding of the making-relatives ceremony to different authors who wrote it down. In order to effect reconciliation between Indigenous groups, they would perform the making-relatives ceremony. They would do a

---

40. See Miller, "Compact, Contract, Covenant," 15–17.

mnemonic ritual to remind themselves of some of the sources of the enmity between them; they would then burn sweetgrass, pray, and would thus make peace and commit themselves to living like relatives. With the signing of the Saskatchewan Treaties 4 and 6, this was the understanding of the elders. We promised to live like relatives. One story has it that Katchishiway one day said to the government official who was son-in-law to the Queen, "How is my brother-in-law today?" Katchishiway's understanding was that if he was a son of the Queen via the treaty, and the government official was a son of the Queen via marriage, then they were relatives.

Indigenous people understood that we must become relatives; because "Canada" was new to the land, the treaty was a way for them to become related to the land by coming into relationship with the original stewards of the land. We were to become like siblings. We would have a nation-to-nation relationship and live within the shared space or place that is now called Canada. Each individual Indigenous people understood this as applying to relationships with one another as well as the newcomers.

The treaty, then, involves more than the words written on the page but establishes a story that takes the land, Creator, and we human beings into account. The treaty story or narrative extends identity back to our ancestors as we live out the law that flowed from our creation story. Identity is extended to include all who will come after us. When Indigenous people made treaty, they were trying to plan for seven generations, so I was present when my ancestors signed the treaty. Henry Morris, treaty commissioner, understood this treaty as a covenant that would last as long as the grass grew and the rivers ran. We were to become one people.[41]

It is a shared narrative because it is the story of the newcomers as well. It means that newcomers must begin to assimilate a relationship with land and the First Peoples, but the treaty ensures or presses us to make sure that we do not erase the differences between us. For it is the distance that we call honor and respect, and it is the space of collaboration, if we could understand the diversity without becoming suspicious of it. This means that we need greater emotional intelligence to be able to feel in a way that leads to healing and not greater animosity.

This is a shared narrative that could indeed be Canada's creation story that tells how Canadians became related to the earth. Canada's foundation then is not about Canada being a colonial nation-state but about how we came together and continue to move forward in reconciliation to heal the land.

---

41. See the discussion in Morris, *The Treaties of Canada*.

■ Conclusion

We need a new narrative that is built upon the historic relationship that is still active and growing, and that sees us as moving together like relatives within this good world where the Creator has placed us.

It was my hope that newcomers in Canada could be introduced early on to relationships with Indigenous people who would help them to develop and grow a relationship with this land that is called Canada. In this way, the treaty may become a shared narrative that tells Canada and Canadians about how they came to be related to the earth. At the same time, it does not erase the diversity in a form of neo-Christendom, colonialism, or tribalism that then lapses into vendetta as a way forward.

The notion of a treaty could serve as a healing device, as it provides a shared narrative that can function as a creation story. On this basis, we could develop a spirituality and theology of the common good for all who share this place called Canada.

■ Bibliography

Altson, Renée. *Stumbling toward Faith: My Longing to Heal from the Evil That God Allowed.* Grand Rapids: Zondervan, 2004.

Austin, Alvyn, and Jamie S. Scott. *Canadian Missionaries, Indigenous Peoples: Representing Religion at Home and Abroad.* Toronto and Buffalo: University of Toronto Press, 2005.

Barth, Karl. *Church Dogmatics: The Doctrine of Reconciliation.* Edited by G. W. Bromiley and T. F. Torrance. Vol. IV. London: T&T Clark, 2010.

Black, Elk, and Joseph Epes Brown. *The Sacred Pipe: Black Elk's Account of the Seven Rites of the Oglala Sioux.* Penguin Metaphysical Library. Baltimore: Penguin, 1971.

Borrows, John. *Recovering Canada: The Resurgence of Indigenous Law.* Toronto: University of Toronto Press, 2002.

Brokenleg, Martin. "The Spirituality of Self-Determination." Anglican Indigenous Sacred Circle, Prince George, British Columbia, August 7, 2018.

Colwell, John. *Promise and Presence: An Exploration of Sacramental Theology.* Milton Keynes: Paternoster, 2005.

Deloria, Vine, Jr. "Philosophy and the Tribal Peoples." In *American Indian Thought*, edited by Anne Waters, 3–11. Maldan: Blackwells, 2004.

Ellul, Jacques. *The Technological Society.* Translated by John Wilkinson. New York: Random House, 1964. Originally published in French as *La Technique ou Tenjeu du siecle*, 1954 by Max Leclerc et Cie.

Grenz, Stanley J. *Theology for the Community of God.* Grand Rapids: Eerdmans, 2000.

Harper, Stephen. *The Call of the North—Address by the Prime Minister Stephen Harper.* Ottawa: Goverment of Canada, 2006.

Kovach, Margaret. *Indigenous Methodologies: Characteristics, Conversations and Contexts.* Toronto: University of Toronto Press, 2009.

Mackey, Eva. "The Apologizers' Apology." In *Reconciling Canada: Critical Perspectives on the Culture of Redress*, edited by Jennifer Henderson and Pauline Wakeham, 47–62. Toronto, Buffalo and London: University of Toronto Press, 2013.

McLeod, Neal. *Cree Narrative Memory: From Treaties to Contempory Times.* Saskatoon: Purich, 2007.

Meili, Dianne. *Those Who Know: Profiles of Alberta's Native Elders.* Edmonton: Newest Press, 1991.

Miller, J.R. "Compact, Contract, Covenant: The Evolution of Indian Treaty-Making." In *New Histories for Old: Changing Perspectives on Canada's Native Pasts*, edited by Theodore Binnema and Susan Neylan, 66–91. Vancouver: UBC Press, 2007.

———. *Residential Schools and Reconciliation: Canada Confronts Its History*. Toronto and Buffalo: University of Toronto Press, 2017.

Morris, Alexander. *The Treaties of Canada with the Indians of Manitoba and the North-West Territories: Including the Negotiations on Which They Were Based, and Other Information Relating Thereto.* Toronto: Belfords, Clarke, 1880; repr., 2014.

Newbigin, Lesslie. *Foolishness to the Greeks: The Gospel and Western Culture.* Grand Rapids: Eerdmans, 1986.

Preston, Richard J. *Cree Narrative: Expressing the Personal Meanings of Events.* Carleton Library Series. 2nd ed. Montreal: McGill–Queen's University Press, 2002.

Ricoeur, Paul. *Political and Social Essays.* Athens: Ohio University Press, 1975.

Schouls, Timothy A. *Shifting Boundaries: Aboriginal Identity, Pluralist Theory, and the Politics of Self-Government.* Vancouver: University of British Columbia Press, 2003.

Sen, Amartya. *Identity and Violence: The Illusion of Destiny.* Issues of Our Time. 1st ed. New York: W.W. Norton & Co., 2006.

Todorov, Tzvetan, and Andrew Brown. *The Inner Enemies of Democracy.* Cambridge: Polity, 2014.

Truth and Reconciliation Commission of Canada (TRCA). *Calls to Action.* Winnipeg: TRCA, 2015.

Willard, Dallas, Kevin Harney, and Sherry Harney. *The Divine Conspiracy: Jesus' Master Class for Life: Participant's Guide.* Grand Rapids: Zondervan, 2010.

Wilson, Shawn. *Research Is Ceremony: Indigenous Research Methods.* Winnipeg: Fernwood, 2008.

From an Island of Africa to an Island of Europe: Perspectives on a Theology of Creation from Madagascar and Crete

Louk Andrianos Andriantiatsaholiniaina[1]

■ Introduction

Being born and raised in Madagascar, the world's fourth-largest island in the Indian Ocean, next to the African continent, I am now living in Chania on the island of Crete, which is the largest island in Greece and the fifth-largest island in the Mediterranean Sea. A stunning diversity of plant and animal species found nowhere else evolved after the island of Madagascar

---

1. Louk Andrianos Andriantiatsaholiniaina is a World Council of Churches consultant for the care for creation, climate justice and sustainability. He is registered as a coresearcher at the University of the Western Cape for the project on "An Earthed Faith: Telling the Story amid the 'Anthropocene.'"

How to cite: Andriantiatsaholiniaina, LA 2024, "From an Island of Africa to an Island of Europe: Perspectives on a Theology of Creation from Madagascar and Crete", in EM Conradie & WJ Jennings (eds.), *The Place of Story and the Story of Place*, in An Earthed Faith: Telling the Story amid the "Anthropocene", vol. 3, AOSIS Books, Cape Town, pp. 41–58. https://doi.org/10.4102/aosis.2024.BK355.03

broke away from the African continent 165 million years ago.² Similarly, but on a smaller scale, the island of Crete is a biodiversity hot spot, having 223 endemic vascular plants as a result of its long isolation and the wide range of habitats it includes.³

Having lived my youth on an African island and continuing my quest for scientific knowledge on a European island, I was fortunate to experience the spiritual wisdom of the people living in these two island sanctuaries. I am thankful to God the Creator for being able to contribute to the story of these places and place these stories in discourse on creation theology—that is the purpose of this volume.

Time, simply counted with years, vanishes like smoke if there are no histories and records of places. I lived the first 20 years of my life in the place called "the thousand cities," or Antananarivo, on the island of Madagascar and the second 20 years in a place called Chania or Khania, on another beautiful but smaller island of the Mediterranean Sea, namely Crete.

I am delighted to be able to study ecology and theology in practice in Crete, and I am grateful to be able to work at the Institute for Theology and Ecology at the Orthodox Academy of Crete.

This essay relates my engagements in creation theology through stories from these two islands. Starting with the story of the Blue Cross movement in Antananarivo and then continuing with the story of a mirror at the Orthodox Academy of Crete in Kolymvari, this essay ends with the story of my experience in the World Council of Churches (WCC), working in the field of creation care. In this way, I will express my understanding of creation theology on the basis of working with the WCC in the field of economic and ecological justice. This essay also offers some narratives on praising the Creator God for the miraculous way of blessing a place through the histories of faithful believers. It is worth studying the places where this history begins, develops, and ends for the purpose of the wellness of our common home. Here is the history of my place and the place of my history, which began in Madagascar and continues on Crete, with the aim of sharing the good news to the ends of the earth. As it is written in Matthew 24:14: "And this good news of the kingdom will be proclaimed throughout the world, as a testimony to all the nations, and then the end will come" (NRSVue).

---

2. See Dresch et al., "Madagascar."

3. See Menteli, "Endemic Plants of Crete."

■ Sacredness and Virtue in God's Creation in the Malagasy Context

In Madagascar, the majority of citizens are indigenous, with a widespread belief in a unique God who has created everything. The Creator of all is a ruling spirit who is present more intensely in some parts of creation. Such a part of creation may be human, animal, or plant, to which a notion of sacredness should be attributed. A sacred human, mostly an older man or woman (called *olona mpanondro masina* or *mpimasy*—meaning foreteller or oracle) has the role of providing counseling and healing to the community. Sacred animals such as lemurs, or sacred trees such as the *hasina* trees or dragon trees, *kily* or tamarind trees, *Adansonia digitata* or baobab trees are proofs of an incipient theology of creation among Indigenous Malagasy communities.[4]

In Malagasy belief, the presence of God in some parts of creation renders a special power for healing to a person. They believe that every part of creation is interconnected, so the presence of one sacred part of the creation in a certain place becomes a source of shared spiritual power to this region and to all human beings living within this region.

The *hasina* tree (or any other sacred tree) is believed to bring the divine power of healing to the place and to human beings in the vicinity of the tree. Sacred trees can only grow in a sacred place, which is then considered to be the dwelling place of the holy people, the heirs of leaders in ancient society. The question is whether the sacred trees give holiness to the place or whether the place gives sanctity to the plant. In the Malagasy tradition, the answer is certain: the presence of the sacred tree gives sacred power to the place. If a sacred plant cannot grow in a place, it means that something "bad" is occurring in that place. "Something bad" refers to human practices that are contradictory to the way of God. These are sinful deeds or misbehaviors that require repentance and reconciliation with God.[5]

Holy people called *mpanandro* [foreteller], *mpimasy* [oracle], or *olona masina* [sacred people] play a vital role in connecting the rest of creation to God the Creator. Such persons have a divine virtue that is an invisible power. Examples of holy people include princes, astrologers, and diviners. Some creatures, such as mermaids, rare indigenous plants, and sacred idols, who can influence the lives of people, also embody such a divine virtue. The source of this power is the Creator God (*Andriananahary*) and

---

4. See Voahangy Rajaonah, "The Sacred Trees of Madagascar," UNESCO Digital Library, last accessed March 9, 2023, https://unesdoc.unesco.org/ark:/48223/pf0000086063.

5. See Wikipedia, 2020, "Hasina (hery tsy hita maso)," last modified October 6, 2020, https://mg.wikipedia.org/wiki/Hasina_(hery_tsy_hita_maso).

the ancestors. The word *Andriananahary* means the prince (or *Andriana*) who created (the verb *nahary* in the Malagasy language is the equivalent of the verb "to create") everything else. Malagasy people were described as ancestor worshippers by early Christian missionaries from Europe. The Indigenous people in Madagascar believe that the ancestors are active collaborators with God in controlling the well-being of every part of creation. The spirits of ancestors are alive and should be respected if one wants peace and prosperity.

According to research[6], approximately 40% of the populace adheres to traditional religious practices centered around ancestor veneration. This involves burying the deceased in tombs, with a belief that they can influence the lives of the living through rewards or punishments. The paramount deity in this tradition is referred to as *Zanahary* (the Creator) or *Andriamanitra* (the Fragrant One). Local spirits are also acknowledged, accompanied by an intricate network of taboos governing traditional Malagasy life.[7] Around half of the population follows Christianity, with over a quarter identifying as Protestant and roughly one-fifth as Roman Catholic. Notwithstanding the widespread adoption of Christianity, traditional religious rituals, particularly those related to the deceased, continue to be practised. In the northwest, there exists a community of Sunni Muslims.[8]

■ The Origin of Human Virtue in Malagasy Belief

According to the Malagasy tradition on priesthoods, the virtues of princes or priests of creation stem from the children of God, not like the virtues found in other creatures. In ancient times, the sons of God came down to play on Earth, and they did not return to heaven because they sinned by following human customs. That is why the sons of God came to live and be served by humans on Earth. These sons of God took wives and gave birth to human beings who inherited their dignity. These descendants of God's sons were highly respected, served and worshipped by humans because they were precious. These generations are the primary source of nobility. And this is why the ancients say that virtue is what makes a prince noble.

---

6. See Britannica, "Ancestor Worship," last accessed February 25, 2023, https://www.britannica.com/topic/ancestor-worship

7. See Britannica, "Ancestor Worship," last accessed February 25, 2023, https://www.britannica.com/topic/ancestor-worship.

8. See Britannica, "Ancestor Worship," last accessed February 25, 2023, https://www.britannica.com/topic/ancestor-worship.

It is possible to weaken or lose dignity if one commits a forbidden sin, and this is what the ancients call the "death of dignity" or "loss of dignity." Someone who has a sacred virtue should be strict with his or her dignity by abstaining from many things. For example, such sacred people could not marry except with a person gifted with a similar dignity, so that the dignity of the son of God continues to stay on him and does not cease. On the first Thursday of the month, the ancients clean themselves with water that is not touched by dogs and forgive each other in order to restore, revive, and increase their dignity. When a holy person goes to reconcile with other people or sacred things, they use the ritual greeting, "*masina ianareo*," meaning, "may you be holy."[9] In Malagasy belief, any virtue attributed to a human or nonhuman entity always has its origin in the connection with the children of God or *Andriamanitra*.

■ Stories of *Fihavanana*—the Malagasy *Ubuntu*—and the Creation of the Island

Fihavanana is the Malagasy word for a relationship between friends. Friendship, in the Malagasy understanding, could exist between people of any kind of kinship or between any of God's creatures. *Fihavanana* is more than friendship because it has a deep spiritual meaning in the sense of the interdependence between creatures. *Fihavanana* is not for benefits, as its value exceeds any economic value. It is about mutual respect and solidarity between members of the global family of life.

Madagascar does not form part of mainland Africa but, geographically, it belongs to Africa. There are two histories of the creation of the island of Madagascar, following the mythologies shared in scientific and Indigenous communities. The first history is that the island of Madagascar was detached from India, and after a big earthquake it descended to come close to Africa. That theory is justified by the similarity between the physical appearance of the Asian people from Malaysia and the majority of Indigenous Malagasy. Another history has it that the island was separated from the African continent, as can be imagined from the shape of the east shore of the African continent, which corresponds with the shape of the west shore of Madagascar. Also, the African characteristics of inhabitants in the southern part of Madagascar seem to confirm this theory.

Is Madagascar then an island or a continent? Here is one answer to this perplexing question:

---

9. See Wikipedia, 2020, "Hasina (hery tsy hita maso)," last modified October 6, 2020, https://mg.wikipedia.org/wiki/Hasina_(hery_tsy_hita_maso).

Neither geologists nor biologists have a definition that is capable of classifying Madagascar unambiguously as an island or a continent; nor can they incorporate Malagasy natural history into a single model rooted in Africa or Asia. Madagascar is a microcosm of the larger continents, with a rock record that spans more than 3000 million years (Ma), during which it has been united episodically with, and divorced from, Asian and African connections. This is reflected in its Precambrian history of deep crustal tectonics and a Phanerozoic history of biodiversity that fluctuated between cosmopolitanism and parochialism. Both vicariance and dispersal events over the past 90 Ma have blended a unique endemism on Madagascar, now in decline following rapid extinctions that started about 2000 years ago.

Madagascar originated as part of the Gondwana supercontinent. Its west coast was formed when Africa broke off from Gondwana around 165 million years ago. Madagascar eventually broke off from India about 88 million years ago. It is geologically located within the Somali plate.[10]

Madagascar has eighteen tribes, but they all speak the same Malagasy language[11]. One spiritual belief that all Malagasy people firmly hold is the respect of *fihavanana* or "kinship" between all creatures. God created all living beings like one big family. They all have relationships and should love one another. They must live in harmony between themselves and with the rest of creation (plants, animals, rivers, oceans, lakes, mountains, etc.) to enjoy the harmony and beauty of creation. Similar to the notion of *ubuntu* in many African countries, *fihavanana* expresses the importance of respect between all parts of creation. *Ubuntu* is a Nguni term meaning "humanity."[12] It is sometimes translated as "I am because we are" (also "I am because you are"),[13] or "humanity toward others" [isiZulu: *umuntu ngumuntu ngabantu*]. In isiXhosa, the latter term is used, but it is often meant in a more philosophical sense to mean "the belief in a universal bond of sharing that connects all humanity."[14]

The *fihavanana* principle places the benefits derived from relationships above any other benefits, especially economic ones. In Madagascar, an Indigenous theology of creation has its basis in *fihavanana*. For the Malagasy people, God created every part of creation to enjoy the relationship with the others and with the Creator. Animals and plants are

---

10. See De Wit, "Madagascar: Heads It's a Continent, Tails It's an Island."

11. See Britannica, "Madagascar, Ethnic Group," last accessed February 25, 2023, https://www.britannica.com/place/Madagascar/Ethnic-groups

12. Tutu, "Who we are".

13. See Nkem Ifejika, "The question: What does ubuntu really mean?", *The Guardian* (London), September 29, 2006, https://www.theguardian.com/theguardian/2006/sep/29/features11.g2.

14. See DBpedia, 2021, "Ubuntu philosophy," last accessed March 6, 2023, https://dbpedia.org/page/Ubuntu_philosophy.

friends and they are bound by a *fihavanana* spirituality. All creation is blessed when *fihavanana* is respected. In Madagascar, *fihavanana* or "kinship" is the most precious value in life. *Ubuntu* and *fihavanana* also give importance to spiritual interdependence and existence beyond death. Every creature continues to live, even after death, by the continuation of its spiritual existence, so every creature deserves unlimited respect. God the Creator is eternal, and every creature should find as its divine destiny immortality or eternal perfection. That is also found in the Orthodox theology of creation through the concept of the deification[15] of human beings through Jesus Christ and the eschatological consummation of the rest of creation (see Isa 11).

■ A Story of the Youth "Blue Cross" Movement to Protect Creation

In Antananarivo, there was an ecclesiastical movement called Blue Cross, which was led by faithful women of the Fiangonan' i Jesoa Kristy eto Madagasikara (FJKM churches) in Antananarivo. Antananarivo is the capital of Madagascar, and this is the place where I was raised by a family with both a Roman Catholic and a Protestant background. My late mother, Raneny Celine, was one of the restless promoters of the Blue Cross movement, and I used to follow her on a pilgrimage to remote villages around the capital region of Antananarivo. The vocation of the Blue Cross movement is to convince people to protect their own bodies and their families against alcohol abuse. Their understanding of creation theology is to protect human bodies, as a divine creation of God, from any source of distortion or destruction caused by human vices, such as alcohol dependence. Because of a high poverty rate, many Malagasy people seek relief in alcohol. The country produces a cheap alcoholic drink called *toaka gasy* (rum or *rhum* in French), which is produced from the artisanal distillation of sugar cane juice. Because of its high and uncontrolled alcohol by volume (around 40%–52%), Malagasy *rhum* can destroy human brains and lungs, and thus the overall health of the individual and, by extension, the mental health of whole families. It makes couples divorce and leaves children without parental care. When a person is drunk on *rhum*, he or she becomes a source of blasphemy and causes the destruction of God's creation. A drunk person is capable of doing every kind of harm to any part of creation. A Malagasy saying holds that a drunk person can burn a forest or kill sacred animals without realizing it.

---

15. Russell, *The Doctrine of Deification*.

It is true that alcohol production from sugar cane is a great source of revenue for the island of Madagascar. Alcohol is useful for many purposes, such as in medicine as a disinfectant or in cosmetics as a solvent. In developed countries, expensive beverages based on alcohol, such as whiskeys, are highly valued. In poor countries, cheap alcohol such as *rhum* is used as an easy way to forget the pains and poverty of daily life. Alcohol attacks the nervous system of the drunkard and it changes their relation with the external environment in a radical way. Alcohol abuse is a sin against creation. It could lead to a whole house or even a forest burning down. Excessive alcoholism is an example of human greed to satisfy an unlimited desire without considering the well-being of the other parts of creation. God created sugar cane plants and human creativity, but the desire of humans for more can lead to the destruction of many parts of creation if it is not limited by sound spirituality, namely the fear of God.

In Madagascar, many children and families have suffered because of the abduction of fathers or mothers by *rhum*. The government has no organized plan to help these people, so local churches are trying to offer spiritual and social assistance to the victims. Many claim that alcoholic drinks are not bad in themselves, but we all admit that the uncontrolled overuse of and addiction to alcohol may transform human beings into troublemakers. One of the elders in our local church told us a sad story of a boy who used to drink *rhum*, and one night he was so drunk that he killed his mother without knowing that he did that. When the boy woke up the following day, he cried and regretted it bitterly because he did not realize what he had done, but by then it was too late.

The Blue Cross movement is based on Christian teaching, which holds that human bodies should be the temples of God's Spirit, and they should not be destroyed by alcoholism or sinful human desire. The movement is an act of advocacy and care for creation on the island of Madagascar. In order to reach remote villages outside the city of Antananarivo, female members of the Blue Cross movement made pilgrimages, and they usually carried bags loaded with leaflets and New Testaments for distribution. During their journey, they shared kind words and life-transforming stories with youth and villagers. Most of these women have their own stories about the effects of alcoholism on their lives. Some argued that excessive alcoholism is a weapon used by the governing powers to keep workers in a vicious cycle of greed and poverty. Fortunately, the teaching of Jesus Christ can help people tackle greed and avoid the slavery of alcoholism (Luke 12:5). According to Luke 12, Jesus said that life does not consist of what we have to eat or drink only. People do not live only by bread but by every word of God. The Blue Cross movement proclaims Jesus's teaching by preaching and singing hymns dedicated to creation care, starting with oneself. For youth followers of the Blue Cross movement, there were some alternative advocacy activities involving sports and choirs in the church. For example, a Blue Cross basketball team and a choir called Fanantenana (meaning "hope"), or the Blue Cross Choir, were created. A dozen young kids belonging to the Blue Cross movement used to meet every week to praise God the Creator with new hymns and activities in the local church of Ambondrona to which I belonged.

By God's grace, I studied environmental sciences at the University School for Agronomy in Antananarivo. My interests were in medicinal plants and biodiversity conservation. The idea of healing creation and helping people through environmental care pushed my quest for more scientific knowledge. In parallel, I was very keen on studying the Bible, without aiming to become an ordained priest. Instead, my aim was simply to act as a lay scientist and ecotheologian whose concern is to serve God the Creator and creation as a whole.

After finishing my university studies in Madagascar, I earned a French scholarship to study hydrology in Belgium, more precisely at the Free University of Brussels. Brussels is a cosmopolitan and multicultural city, where people learn to live together harmoniously, especially Christians and Muslims. Because of my passion for studying the Bible and the Greek language, I continued my studies in Greece, which is considered to be a holy land for Malagasy Christians. The Greek language, in which the New Testament was originally written, is also considered to be a divine language. Here, an assertion of Malagasy creation theology comes into play, namely that the history of a place makes this place holy. I am thankful to the Greek State Foundation (*Idrima Kratikon Ypotrofion*, IKY) that offers gives scholarships to Malagasy students for doctoral studies in Greece. The topic of my research was sustainability and environmental management, which was a prototype subject at the Technical University of Crete in the 1990s.

■ A Story of the "Mirror" at the Orthodox Academy of Crete to Praise the Creator[16]

Scientists explain the formation of the island of Crete as follows:

> 20 million years ago, the two tectonic plates holding Africa and Asia crashed together. By this point, the land that we recognize as Crete had completely emerged from the Tethys Sea, close to the crash point of the African and Asian tectonic plates. More than 500 million years ago, the area of Crete was submerged in Tethys Sea and life was only marine. The sediments moved by the rivers of the coasts of Pangaea and the wind, gathered and mixed with the shells of the

---

16. the following section is based on the following source: Semon Central, n.d. "Preaching Ideas & Sermon on Matthew 5:14," last accessed November 24, 2023, https://www.sermoncentral.com/sermon-illustrations/on-scripture/matthew-5-14?keyword=Matthew+5%3A14.

sea organisms, forming layers of rocks. About 200 million years ago Pangaea began to break apart, forming the continental plates. 70 to 55 million years ago the plates of Africa, India and Lavrasia crashed violently and forced the bottom of Tethys to raise and form a chain of mountainous folds and sinkings from Atlas and the Pyrenees to the Alps and further to Caucasus and the Himalayas. A part of this chain reaction took place in the Balkans and the Minor Asia. This is when the Dinarotauric Arrow started forming. This mountainous arrow begins at the Dinaric Alps, runs along the range of Haemus, crosses the Greek territory and ends at the Range of Taurus in Minor Asia. Current Crete was submerged at the middle of this arch.[17]

Crete was known from the biblical verse in Acts, mentioning the wreckage of the ship with the Apostle Paul in the south of the island (Acts 27:13). All Cretans know about this history, and they call this place *kali limeni* (meaning "good harbor"), where Saint Paul performed a miracle in the name of Jesus Christ among the people of Crete. Personally, I realized that the name found in the Bible, identified with the island of Crete, is the "isle of Caphtor," and the people from Crete are the Cherethites (2 Sam 8:18; Jer 47:4; Gen 10:14; Deut 2:23; Amos 9:7; 1 Chr 1:12).

To better understand the history of the "mirror" at the Orthodox Academy of Crete, one should refer to the descriptions from the book entitled *It Was on Fire When I Lay Down on It*,[18] written by the American author Robert Fulghum. In this book, he tells how he found purpose in life after his visit to the Orthodox Academy of Crete. A preacher from a Lutheran church used the story of Fulghum and the mirror, describing to his congregation how Fulghum took a course on Greek culture at an Orthodox retreat center on the island of Crete. A man named Alexander Papaderos had built the center. After World War II, Papaderos became disturbed by the hatred that his people still had for the Germans. In response, he built a meeting place where people could come to make peace, talk, and try to understand one another. The center was built on the site where Nazi soldiers had brutally murdered thousands of Cretan civilians. For years, people had come from all over the world to share in the love and grace of Dr. Papaderos and to learn a better way.

Fulghum raised his hand and asked, "Dr. Papaderos, what's the meaning of life?" The class chuckled, but Papaderos nodded and said, "I will answer your question." Taking his billfold out of his pocket, he brought out a small round mirror, about the size of a quarter. He spoke about growing up poor in a small village in Crete. One day, during the war, a German motorcycle was wrecked near his home, and Papaderos picked up a piece of the

---

17. See "Crete's Geology: How was it formed?", Cretan Beaches, last accessed February 22, 2023, https://www.cretanbeaches.com/en/facts/geology-of-crete.

18. Adapted from Fulghum, *It Was on Fire When I Lay Down on It*, 170–75.

broken mirror from the motorcycle. He scratched it on a stone to round off the edges, and then began to play with the mirror as a toy. He became fascinated by the way that he could use the mirror to reflect sunlight into places where light would never shine. It became a game to get the light into dark places.

As he became a man, Papaderos began to realize that this was more than a child's game; it was a metaphor for what he wanted to do with his life. He understood that he was not the source of the light, but the light was there already—the light of truth, understanding, knowledge—and it would only shine in dark places if he reflected it.

Papaderos told the class, "I am a fragment of a mirror whose whole design and shape I do not know. Nevertheless, with what I have, I can reflect light into the dark places of this world—into the black places in [our hearts]—and change some things in some people. Perhaps others may see and do likewise. This is what I am about. This is the meaning of my life."[19]

As it was experienced by Cretan people, a theology of creation in Orthodoxy is closely related to the meaning of life. God created human beings and the rest of creation for a divine purpose. Human beings are the priests of creation. Their role is to care for creation and present it for God's glory toward the eschatological deification of all. The history of the miraculous healing of Saint Paul from the venomous snake, along with the history of the mirror of Dr. Papaderos, demonstrate the mighty dependence of all creation's activities on the divine control of the Creator God, called Παντοκράτορας [Almighty Pantocrator] in Orthodoxy, meaning the One who holds in control everything in every time with mighty power. The key to the fullness of life and wholeness of all creation is to accept the light of Jesus Christ, who can save every part of creation from eternal destruction. From the time of birth to the time of dormition of a living being, God is giving the chance of deification to every creature. Deification means reconciliation and unity with God in the earthly and heavenly life after dormition.

■ Feminine Influence in Understanding an Orthodox Theology of Creation

To use the Orthodox term "dormition" or κοιμηση for the physical "death" of a believer in Christ, the dormition of my mother on the island of Crete has helped my understanding of creation theology in the Eastern Orthodox tradition. The secularized world teaches people to behave according to

---

19. See Scott J. Higgins, "Alexander Papaderous—Reflecting Light into Dark Places", November 28, 2012, https://scottjhiggins.com/alexander-papaderous-reflecting-light-into-dark-places/.

geographical or cultural standards; as a result, everyone cares only for their own interests within territorial borders and religious denominations. God created the whole universe, and God cares for every single part without borders or any kind of discrimination. God sent to me another woman to accompany me on my earthly journey. Her name is Georgia, and she is from Crete. We married almost one year after my mother passed. My shared life with Georgia marks the beginning of my integration into the Greek culture and Orthodox tradition. God has a mysterious plan for every creature, and I realize that my new life as a Greek citizen is God's divine path for me to get involved with creation theology from an Orthodox perspective. God created men and women to create a family, to multiply, as it is written in Genesis, and to continue God's creation. Christian teaching instructed me that "life is more important than clothes," so I considered that the unity of my new family is much more important than my career. After almost two years of experiencing North American culture, we decided to return to Greece to live on Crete permanently in 2002. In Greek culture, the concept of family unity is sacred. According to Orthodox creation theology, men were created to build families with women. They should support one another to raise their children in close relationship with other members of the extended family. The concept of divorce between spouses or dispute among relatives is associated with spiritual sin, which is condemned by the Creator God. It is common to find a Greek family where adult children—at the age of 30 years or more—continue to live with their parents because they believe that is the right thing to do in the eyes of God. Besides economic circumstances, this Orthodox concept of family may be contrasted with what is happening in Western European culture, where parents send their children outside the family home at the age of nineteen years (Sweden) or 25–27 years (the average in Europe).[20]

Life in Crete is not easy for a person like me because it is difficult to find an academic job there. I worked at a private company in Rethymnon before I saw in the newspaper that the Orthodox Academy of Crete was looking for a scientist with a theological background to lead its Institute for Theology and Ecology in Kolymvari. At that time, corruption was widespread in Greece, so many of my friends advised me that my chances were very small. I should have had a supporter from within or from someone renowned in political or religious leadership to get such a position. I prayed, saying to myself that I have the support of the highest priest and savior of all creation, who is Jesus Christ! He died on the cross and has been resurrected for the salvation of the whole of creation by faith. A few days after my application,

---

20. See Eurostat, "Archive: Age of young people leaving their parental household," August 6, 2020, https://ec.europa.eu/eurostat/statistics-explained/index.php?oldid=494351. See also www.iariw.org.

Dr. Papaderos, the General Director and co-founder of the Orthodox Academy of Crete (OAC), called me for an interview. When I faced him, I felt the light from his "mirror" and was convinced immediately that we shared the same light of Jesus Christ, who called us to care for creation. There were many applications for the position, but he decided to introduce me to the board of the OAC as the most suitable candidate. Archbishop Irineos, the founder of the OAC, was the president of the OAC at that time, and I remember having seen him once when Georgia, now my wife, took me to his office to ask for the blessing of our relationship. Usually, no parents would allow a Greek girl to get married to a boy from a religious tradition other than Orthodox. Archbishop Irineos was firm in encouraging us that we are all children of God, and the faith in Jesus Christ covers all laws and traditions. He blessed us and advised us to keep the "love" which is God: "God is love, and all who live in love live in God, and God lives in them" (1 John 4:16).

■ Living a Theology of Creation with the Ecumenical Patriarchate of Constantinople and the World Council of Churches

In 2006, I started working at the OAC to lead the Institute for Theology and Ecology (ITHE) in Kolymvari, where I met Ecumenical Patriarch Bartholomew, who is the spiritual mentor, founder, and adviser of the institute since its establishment in the 1990s. The Ecumenical Patriarch is known as the "Green Patriarch" because of his actions and advocacy for the protection of God's creation worldwide.

Among others, the organization of interdisciplinary meetings on ecotheology was one important part of my responsibilities at the OAC. At that moment, the Foundation for Research and Technology, Hellas (FORTH), based in Heraklion, collaborated with our institute, and we managed to establish a biannual conference on Ecological, Theological, and Environmental Ethics (ECOTHEE), which started in 2008. Since then, ECOTHEE conferences gathered theologians, scientists, politicians, priests, and other lay or religious people from all over the world. The ECOTHEE conferences are always organized under the auspices of the Ecumenical Patriarch Bartholomew I, and we often host participants from interfaith and ecumenical organizations. The village of Kolymvari, the place where the ECOTHEE conferences are hosted, became a reference center of blessings from Constantinople, from which the Ecumenical Patriarch sent a formal blessing in the form of an opening letter to each ECOTHEE conference. Without the spiritual guidance of the Green Patriarch, the development of such theological work for the sake of creation would be very restricted and difficult.

The Green Patriarch, His All-Holiness Ecumenical Patriarch Bartholomew,[21] was the first religious leader who stated that any harm to the environment is a sin against God and creation. This is an important element of creation theology in the Orthodox perspective. Every action or status of any part of creation has a direct relation with God and other creatures. God created the whole of creation in perfect harmony and intrinsic interdependence for God's glory in Christ. Any part of creation, especially human beings, should boast of their beauty or strength or craftiness. All is God's and for God's creation. To address the global environmental crisis, the ITHE strives to practice this kind of creation theology by building an environmental ethos for current societies. Climate change, biodiversity loss, pollution and destruction are signs pointing toward an urgent call to action and repentance for ecological conversion. We are living in a critical time to save creation from irreversible destruction. This is a *Kairos* for eco-conversion, as it was described following an international conference on ecotheology in Wuppertal, Germany.[22]

Given such a *Kairos*, the institute committed itself to accelerate its work on ecological justice and to engage in ecumenical and interfaith collaboration through various activities. One of our responses is my initiative to organize the movement toward a Sustainable Alternative for Poverty and Ecojustice (SAPREJ), which is a kind of pilgrimage toward justice and peace for God's creation in poor countries. It gathered theologians, scientists, and lay people from Christian, Muslim, Buddhist, and Indigenous communities to discuss and act together for the protection of the marginalized natural world and vulnerable people. SAPREJ started with conferences in Kolymvari in 2012 and then continued in Madagascar in 2016, then in Uganda in 2018, in Fiji in 2020, and recently in Ethiopia in 2023. It is amazing to see the hand of God in moving people's hearts to care for creation. We stated that the root of the ecological crisis is human greed, and the solution lies in the repentance of all stakeholders in society. Not only politicians and scientists but all, especially the religious communities, ordinary citizens, and Indigenous people.

His All-Holiness Ecumenical Patriarch Bartholomew said, "As Orthodox Christians, we use the Greek word 'Kairos' to describe a critical moment in time, often a specific historical period with lasting repercussions and eternal significance. For humankind and the planet as a whole, now is our Kairos: the decisive time in our relationship with all of God's creation, when we must respond in an opportune manner to protect life on Earth from the worst

---

21. See Greek Orthodox Archdiocese of America, "The Green Patriarch: His All-Holiness Ecumenical Patriarch Bartholomew," last accessed February 15, 2023, https://www.goarch.org/society/greening-the-parish/green-patriarch.

22. See Andrianos et al., *Kairos for Creation*.

consequences of human recklessness. May God grant us the wisdom to act promptly."[23]

Following the example of the Green Patriarch, the OAC organized ecological events involving primary and secondary schools at a local level in Crete in collaboration with ecumenical and interfaith partners at the international level. As a result, the WCC noticed our work, and the working group on the greed line came to Kolymvari to meet us and organized their meeting in Crete. The WCC also affirmed that human greed is the root cause of injustices and the destruction of the natural world. The unlimited human desire to have more and more material possessions is the simplest definition of greed. The ancient Greek philosophers had some great insights on greed, saying that happiness is found in moderation. Moderation is the key to controlling greed: "Give us today our daily bread" (Matt 6:11) is the teaching by Jesus Christ on moderation in consumption and the virtue of ascetic life—as is promoted in Orthodox communities and monasteries.

■ The Greed Line Study

The hospitality and the moderate traditional lifestyle of the Cretan people, especially in monasteries, surprise every guest at the OAC. Dr. Papaderos had an explanation for the negative descriptions of Cretans in the Bible, such as in Titus 1:12–13, where Paul wrote about the Cretans who "are always liars, evil brutes, lazy gluttons." He said that these are the characters of the Cretan mercenaries at that time, not the Indigenous Cretans who were victims of colonizers from different neighboring countries. Crete was colonized by the Romans, Ottomans, Arabs, Venetians, and German Nazis, in chronological order.

During a visit of delegates from the WCC to Crete, I presented a study on how to measure the greed line as a tool to promote justice and peace with creation. This study is an extension of my doctoral research in developing a measurement for sustainability. The idea is to define a minimum and maximum value for every indicator of greed or sustainability. Humans, as priests of creation, should operate within planetary limits so as to avoid the destruction of the whole of creation. The exploration of the concept of a greed line continued for several years and ended with the publication of a book by the WCC's ecological and economic justice team.[24]

---

23. See Greek Orthodox Archdiocese of America, "The Green Patriarch: His All-Holiness Ecumenical Patriarch Bartholomew," last accessed February 17, 2023, https://www.goarch.org/society/greening-the-parish/green-patriarch.

24. See WCC, "The Report of the Greed Line Study Group of the WCC," last accessed February 18, 2023, https://www.oikoumene.org/resources/documents/the-report-of-the-greed-line-study-group-of-the-wcc.

I am grateful to God the Creator that I was involved in developing the concept of a greed line so as to protect the sustainability of creation.[25]

Since its tenth general assembly in Busan in 2013, the WCC has continued to work on ecological justice and peace by tackling greed. After the recent eleventh assembly in Karlsrühe, the focus of the WCC is on advocacy for climate justice and biodiversity protection. Part of the WCC program is to develop a theology of creation that could bring changes in the lives of the churches as active caretakers of creation. For many years, the WCC has been involved in high-level international, political, and scientific dialogues on the protection of the environment. Ecumenical and interfaith representatives are present in every Conference of Parties (COP) in terms of the United Nations' Framework Convention on Climate Change (198 parties) following the "Earth Summit," which was held in Rio de Janeiro, Brazil, June 03-14, 1992.[26] The recent involvement of the WCC team at the COP27 on climate change and the COP15 on biodiversity reflects the new narratives of living a theology of creation at an ecumenical level. Almost two decades since my engagement in protecting Madagascar's environment, this has yielded global advocacy for creation justice. From an island to an island, God the Creator paved my way to become an ecumenical and interfaith priest of creation, as is formulated in Orthodox theology.

Over the last four years, I have been actively involved in the ecumenical movement on a "Season of Creation" (SOC), which aims at mobilizing faith-based communities to work together for the care of creation on a planetary level. As a member of the steering committee of the SOC, I meet virtually every two weeks with colleagues from many other denominations, including Roman Catholics, Lutherans, Anglicans, Reformed Christians, Methodists, and faith-based organizations (FBOs), to discuss ways to unite religious communities for the yearly activities of a SOC. Each year, from September 01 to October 04—the formal period for the Season of Creation—the Christian family unites for this worldwide celebration of prayer and action to protect our common home. As followers of Christ from around the globe, everyone embraces the common call to care for creation. All beings are co-creatures and part of all that God has made. Human well-being is interwoven with the well-being of the earth. Believers rejoice in this opportunity to safeguard their common home and every part of it.

---

25. See Peralta and Mshana, *The Greed Line*.

26. See United Nations, "United Nations Conference on Environment and Development, Rio de Janeiro, Brazil, 3-14 June 1992," last accessed February 17, 2023, https://www.un.org/en/conferences/environment/rio1992.

Every year, the SOC defines a theme to focus on with the prayers and activities. For 2023, the theme was "Let justice and peace flow." As the prophet Amos cries out: "But let justice roll down like water and righteousness like an ever-flowing stream!" (Amos 5:24, NRSVue) So believers are called to join the river of justice and peace, to take up climate and ecological justice, and to speak out with and for communities most impacted by climate injustice and the loss of biodiversity.[27] All people of God should work together on behalf of all creation as part of that mighty river of peace and justice.

▊ Conclusions

Both the island of Madagascar and the island of Crete fell victim to colonization and religious oppression. As an African country, Indigenous people in Madagascar kept an ecofriendly spirituality that is similar to the notion of *ubuntu*, which maintains that "I am because you are." In Madagascar, *fihavanana*, or kinship, is the most precious value in life. *Ubuntu* and *fihavanana* give the highest importance to spiritual existence beyond death. Every creation is alive even after death by the continuation of its spiritual existence, so every creation deserves unlimited respect. God the Creator is eternal, and every creature has the divine destiny to reach immortality and perfection. This is described in the Orthodox theology of creation through the hope for the deification of human beings through Jesus Christ. The salvation of the rest of creation is dependent upon the deification of human beings, as is written in Isaiah 11:5-7: "Righteousness shall be the belt around his waist and faithfulness the belt around his loins. The wolf shall live with the lamb; the leopard shall lie down with the kid; the calf and the lion will feed together, and a little child shall lead them. The cow and the bear shall graze; their young shall lie down together; and the lion shall eat straw like the ox" (NRSVue).

Malagasy and Cretans are religious people, especially in the quest for spirituality and exchange with the rest of creation. In Acts 2:11, it is written that "Cretans and Arabs—in our own languages, we hear them speaking about God's deeds of power" (NRSVUE). People from Crete were in Jerusalem when the Holy Spirit came for the first time to the apostles. Afterward, Saint Titus, a disciple of Saint Paul, built the first Christian church on Crete. Titus taught many ecclesiastical and ethical lessons to Cretans, who kept these teachings as part of their moralities from generation to generation.

---

27. See the home page at https://seasonofcreation.org/.

From island to island and through different places, I saw the beauty of creation, which reveals the glory of the Triune Creator. With the concepts of *fihavanana* and deification, from an Indigenous spirituality to Orthodoxy and the ecumenical movement, I conclude my narratives on creation theology through stories of places and the place of stories by affirming that God created every part of creation so as to be respected and to be accountable for taking care of each other in unity and reconciliation with the Almighty Pantocrator, Jesus Christ. For all may be one and have life in full for the glory of the One who created all: "that all of them may be one, Father, just as you are in me and I am in you. May they also be in us so that the world may believe that you have sent me" (John 17:21, NRSVue).

■ Bibliography

Andrianos, Louk, et al., eds. *Kairos for Creation: Confessing Hope for the Earth*. Solingen: Foedus-Verlag, 2019.

De Wit, Maarten J. "Madagascar: Heads It's a Continent, Tails It's an Island." *Annual Review of Earth and Planetary Sciences* 31 (2003), 213–48. https://doi.org/10.1146/annurev.earth.31.100901.141337

Dresch, Jean, et al. "Madagascar", *Encyclopedia Britannica*, August 25, 2022. Last accessed September 29, 2022. https://www.britannica.com/place/Madagascar.

Fulghum Robert. *It Was on Fire When I Lay Down on It.* New York: Ballantine Books, 1991.

Menteli, Viktoria, et al. "Endemic Plants of Crete in Electronic Trade and Wildlife Tourism: Current Patterns and Implications for Conservation." *Journal of Biological Research-Thessaloniki* 26 (2019), 10. https://doi.org/10.1186/s40709-019-0104-z

Peralta, Athena, and Rogate Mshana, eds. *The Greed Line: Tool for a Just Economy.* Geneva: WCC, 2016.

Russell, Norman. *The Doctrine of Deification in the Greek Patristic Tradition*. Oxford Early Christian Studies. Oxford: Oxford University Press, 2006.

Tutu, Desmond. "Who We Are: Human Uniqueness and the African Spirit of Ubuntu" (2013). YouTube. Archived from the original on November 10, 2021.

Ecotheology and Creation Theology: Shona People's Indigenous Cosmologies

Sophia Chirongoma[1]

■ Introduction and Self-Location

Creation stories are pivotal pillars, anchoring us and helping us to answer some of the most fundamental questions of existence. They help us to make sense of our place in society, our place in the story of humankind, and above all, our place in the story of the universe. They help us to grapple with the two key questions raised in this volume: "What difference does it make to the story of cosmic, planetary, human, and cultural evolution to re-describe this as the creative work of God's love? What difference does it make to the story of God's love to describe it in evolutionary terms?"

---

1. Sophia Chirongoma is a senior lecturer in the Religious Studies Department at Midlands State University, Zimbabwe. She is also a research fellow at the Research Institute for Theology and Religion in the College of Human Sciences, University of South Africa. She is registered as a coresearcher at the University of Western Cape for the project on "An Earthed Faith: Telling the Story amid the 'Anthropocene.'"

How to cite: Chirongoma, S 2024, "Ecotheology and Creation Theology: Shona People's Indigenous Cosmologies", in EM Conradie & WJ Jennings (eds.), *The Place of Story and the Story of Place*, in An Earthed Faith: Telling the Story amid the "Anthropocene", vol. 3, AOSIS Books, Cape Town, pp. 59-78. https://doi.org/10.4102/aosis.2024.BK355.04

Among the Shona people in Zimbabwe, besides the biblical creation story in the book of Genesis, two of the widely known myths of creation derived from Indigenous cosmologies are the Mwedzi[2] myth of creation, also known in other versions as the Dzivaguru[3] creation myth and the Guruuswa[4] myth. While acknowledging that these two myths offer profound wisdom in helping the Shona people to trace their origins and to explicate the story of our place, this essay also utilizes ecofeminist lenses to contend that both the Mwedzi/Dzivaguru and Guruuswa creation myths are replete with patriarchal traits, which tend to perpetuate male domination and female subordination. Unpacking the contents of the two myths, I will note that this pattern plays itself out in many other myths of creation emerging from patriarchal societies, not only in Africa but throughout the world. I maintain that while these Indigenous cosmologies acknowledge that the earth belongs to both human and nonhuman creatures, the reality on the ground is that there is a binary between the pedestals on which men and women are placed in a patriarchal society—with men placed on a higher pedestal than women. Similarly, the human species tends to have drifted from its stewardship role by treating the earth as a commodity to be exploited through amassing land and other natural resources. Concurring with ecofeminist theologians, I advocate for a responsible ecological stewardship and a theology of fair and equitable distribution of resources.

Drawing insights from my social location as a Karanga[5] person and an African ecofeminist theologian, I acknowledge that the Shona people, like many other Indigenous communities, have a tripartite cosmological view.[6] Alert to the patriarchal innuendos that shape the Shona worldviews, I foreground the gendered perspectives common in both the Mwedzi/Dzivaguru creation myth and the Guruuswa creation myth, their Indigenous views on land ownership, as well as their Indigenous perspectives on the ecological relationship between human and nonhuman entities. It is from within the Shona people's worldview and my positionality as an African ecofeminist theologian that I explore the interface between ecotheology and creation theology.

---

2. Mwedzi is a Shona word referring to the moon.

3. Dzivaguru is a Shona word meaning the great or deep pool.

4. Guruuswa is a Shona word denoting a place with a lot of long grass.

5. The Karanga people are one of the subgroups of the Shona.

6. See Taringa, "How environmental is African traditional religion?"; Mwandayi, *Death and After-Life Rituals*; Mahohoma, "Experiencing the Sacred"; Chirongoma, "Karanga-Shona Rural Women's Agency"; and Chirongoma, "Where Earth and Water Meet."

The two overarching questions guiding the discussion are:

1. How does the Shona people's cosmology help us to conceptualize planetary, human, and cultural evolution?
2. How does the Shona people's ecological and cosmological purview help us to conceptualize the creative work of God's love?

The first part of the essay discusses the two Shona creation myths mentioned above. It brings to the fore the apparent patriarchal leanings in these two creation myths. The second part of the essay draws on these two myths of creation as a lens for critiquing the inherent gender and social inequalities when it comes to land ownership and distribution of natural resources in Zimbabwe. The third part of the essay discusses how the fangs of the global ecological crisis are also biting the people of Zimbabwe, as is manifested in the recurring droughts, cyclones, and erratic weather patterns, vacillating from extreme heat to extreme cold. In unison with ecofeminist theology, I propose a theology of responsible ecological stewardship and a theology of fair and equitable distribution of resources. The concluding section grapples with the apparent paradox between ecological catastrophes, the entrenched gender and social injustices when it comes to access to and distribution of resources, and the belief that an omnibenevolent Creator God is in charge of the universe. Reflecting on the harsh realities of the effects of the global ecological crisis manifesting in Zimbabwe, the essay foregrounds the question of God's creative work and abundant love. It also grapples with the question of our place in creation.

■ The Mwedzi/Dzivaguru Creation Myth

As noted by *Mavhu F.W. Hargrove,*[7] there are various versions of the Mwedzi/Dzivaguru creation myth among the Shona. According to the commonly cited version of this myth, Mwari (God) created Mwedzi (the moon) in a deep pool of water. Bored with the monotony of residing in water, Mwedzi begged Mwari for permission to live on land. No sooner had Mwedzi settled on land than he started feeling lonely and became restless. Empathizing with Mwedzi, Mwari sent Masasi (the morning star) down to become Mwedzi's wife. Mwari also notified Mwedzi that after two years, he would have to return Masasi to the sky. Masasi stayed with Mwedzi and gave birth to all the vegetation on Earth. After two years, a reluctant Mwedzi sent her back to the sky.

Mwedzi started exhibiting loneliness and restiveness again, thus spurring Mwari to send him Vhenekatsvimborume (the evening star). Once again,

---

7. See Hargrove, "Guest Voices."

Mwari warned that she must return to the sky after two years. As she communed with Mwedzi, Vhenekatsvimborume started bringing forth various life forms. Firstly, she gave birth to herbivores and birds and then to boys and girls. After two years, Mwari asked Mwedzi to return her. However, Mwedzi refused. On the next day, Vhenekatsvimborume gave birth to lions, scorpions, and other predators.

It is important to note the intense impact of the Mwedzi myth on how the Shona view the cosmos. For instance, the belief that the first man, Mwedzi, was created in a deep pool of water has profoundly influenced the Shona people's reverence for natural water bodies. Furthermore, the Mwedzi myth is a constant reminder that among the Shona people, there is no clear dichotomy between our past and our present. Hence, we cannot really demarcate between ourselves, our ancestors, and our deities; we just reflect each other like the moon in a pool of water. Akin to the biblical creation myth (Gen 2:7), the man (Mwedzi) is created first and the woman (Masasi) second.

It is interesting to note that unlike the biblical creation account, whereby Adam and Eve are bequeathed the earth as their abode, in the Mwedzi creation myth, the woman is presented as a being who is only temporarily brought to Earth for procreation purposes. Once Masasi and Vhenekatsvimborume had served their utility (bringing forth life), they were intended to return to the pool they originally came from, and it was the man's responsibility (Mwedzi) to return the woman to the pool. Wearing African feminist lenses, we can see how this elicits a sense of male dominance and female subordination. Firstly, it is the man who precedes the woman in the order of creation. Secondly, it is the man to whom Mwari gives the instruction regarding the tenure of the woman's stay on land, and when the tenure elapses, it is Mwedzi's responsibility to return her to the pool. It is therefore the man's volition to return the woman to the pool, and it appears as if the woman has no voice regarding her fate. As such, Mwedzi took it upon himself to return Masasi to the pool after the stipulated two years. However, when the time came for the second woman, Vhenekatsvimborume, to return to the pool, Mwedzi did not want to let her go; hence, he disregarded Mwari's instruction and decided to extend her stay.

As if to punish both Mwedzi and Vhenekatsvimborume for disobedience, the day after they had extended her return to the pool, she started bearing lions, scorpions, and other predators, life forms that are considered harmful and a menace to humans and other animal species. Just like in the biblical account of the fall of humanity, where the blame for the origin of sin and death is put on Eve's shoulders, Vhenekatsvimborume is presented as the one who birthed the predators which threaten human existence and other life forms. The two women mentioned in the story (Masasi and Vhenekatsvimborume) are presented using a highly patriarchal worldview where women are "seen" and not "heard." Using a patriarchal script, the two women are the epitome of docile and service-oriented beings. The two male beings in the story (Mwari and Mwedzi) deliberate and decide on the women's fate. This echoes the views raised by African feminist theologians who bemoan the subordinate status of women, leading to them being treated as nonentities, with a male suzerain to lord over them.[8]

The Dzivaguru creation myth is another version of the Mwedzi creation myth. As retold by Harold Scheub,[9] the Dzivaguru myth depicts Mwari, the Supreme Being, as the god of fertility, the sower, the rain-giver. Dzivaguru is depicted as a simple man who wore his hair long and wore nothing but a belt around his loins. He wandered about the countryside doing acts of kindness and performing magical wonders. Wherever Dzivaguru went, good rains followed, and the people prospered. Dzivaguru is one of the praise names used for Mwari, because he supplies the people with rain. Dzivaguru is also the name of one of Zimbabwe's most important shrines in the Matopos. According to the Dzivaguru myth, Mwari is both male and female. As a female being, Mwari is immersed in the pool with its darkness and mystery; this is the god of below. As a male being, Mwari is the owner of the skies, the god of light, the father of creation who manifests himself in lightning or the shooting star; this is the god of above. He is an ambivalent god, both immanent and transcendent. He is ever-present in his own creation.[10]

Harold Scheub narrates the Dzivaguru myth of creation as follows:[11]

> Mwari put his creation, Musikavanhu, into a deep sleep and then let him drop from the sky. While he fell, Musikavanhu awoke and, in the distance, saw a white stone which was also dropping from the sky at great speed. Musikavanhu began to fly towards the stone, and the closer he got to it, the bigger the stone became, and finally he could no longer see where it finished on either side. Musikavanhu fell softly onto the stone, and the first spot his feet touched softened and emitted water. This place became the stone of the pool, today called Matopos, a place that is venerated. Musikavanhu, bored, began to wander about. When night fell, he sat down near the stone and slept. In a dream, he saw the birds in the air, and many animals on the earth that were jumping from stone to stone. When Musikavanhu awoke, he was surprised to see that all he had just dreamt had become reality. Mwari told Musikavanhu what he was allowed to eat, and what food was forbidden. He was free to eat vegetables, and fruit from the trees, but not to kill and eat animals. Nor were the animals allowed to eat each other.

---

8. See Kanyoro, "Engendered Communal Theology"; Oduyoye, *Introducing African Women's Theology*; Chirongoma and Zvingowanisei, "Patriarchy as an Archetype of Empire"; and Siwila and Kobo, *Religion, Patriarchy and Empire*.

9. See Scheub, *A Dictionary of African Mythology*, 9–12.

10. See Scheub, *A Dictionary of African Mythology*, 10.

11. Scheub, *A Dictionary of African Mythology*, 11.

One day, while Musikavanhu slept, a snake crept over his loins and left its marks. When he woke up, he was overcome by a strange feeling; he had trouble breathing and his penis moved like a snake. A voice told him to go to the pool, and the pain would pass. On his way there, he saw a beautiful young woman sitting on a stone near the pool. She looked like him, but she could neither speak nor move. Musikavanhu heard a voice; instructing him to touch her with his hand. He did, and the young woman came to life, and a snake moved across her loins, too. She was overcome by the same emotions as Musikavanhu. The voice spoke and told Musikavanhu to be kind to his wife, and to all the animals too.

When Musikavanhu had completed the tasks set by God he had to return to heaven. Before he went, he told his children to observe God's laws, or God would punish them. People lived in peace for a long, long time. One day Musikavanhu's children got drunk and became proud. They told the animals and the other people that God was dead and that one of them would be God. God's voice warned them, but because of their pride, they could no longer hear it. God then became angry; he cursed the earth, and the sea's water became salty, the land dried up and thorns grew. During the rainy season, the rivers swept away many people, and crocodiles appeared in the waters. The sun became hot, and the animals began to eat one another and attack men. And men started killing each other.[12]

The Dzivaguru creation myth is another narrative packed with patriarchal assumptions of the male as superior, with positive traits, while the female is pushed to the periphery as an inferior being, endowed with negative traits. Firstly, although Mwari (the Creator) is portrayed as both male and female, the male traits are described in a positive light, which is starkly different from the negative and inferior feminine attributes. He is described as "the god of light, the father of creation," who reveals himself in the form of lightning or a shooting star. He is also depicted as "the god of above." These masculine traits of Mwari elevate the male gender while perching the feminine traits of the deity on a lower pedestal. This comes out clearly when the narrative describes the feminine attributes of Mwari as being "immersed with darkness and mystery," as well as being designated as "the god of below."

Secondly, the male prototype (Musikavanhu) was the first to be created. It is Musikavanhu who is endowed with traits of fertility, making the land bountiful and prosperous by providing rain. On the other hand, the nameless female progenitor is described as a lifeless form who is discovered by Musikavanhu and only comes to life after the "magical" touch from Musikavanhu. Throughout the narrative, Musikavanhu is the doer, while the unnamed female ancestor remains silent and passive, waiting to be brought to life by Musikavanhu.

---

12. I have cited Harold Scheub verbatim at length because his account succinctly describes the Dzivaguru creation myth.

Thirdly, when Musikavanhu's touch jolts her into becoming a living being, the reader surmises that after the snake moves "across her loins," in a similar fashion as it had to Musikavanhu's loins, she conceives and in due course, brings forth some offspring. However, the narrative is silent about her procreative role; the children are referred to as "Musikavanhu's children." When the time eventually comes for Musikavanhu to return to heaven, the narrative only mentions that he counseled "his children" to obey God's laws. Nothing else is said about the woman after her first encounter with Musikavanhu. In unison with fellow African feminist theologians,[13] my reading of the Dzivaguru creation myth exposes a deliberate and continual silencing, trivializing, and domination of women.

An intriguing aspect in this narrative is that the cause of humanity's fall and the depravity of ecology is not blamed on the woman's disobedience but on the disobedience of Musikavanhu's children. Similar to the biblical fall of humanity, disobedience on humanity's part elicits ecological degradation and destruction. This resonates with ecofeminist views that the travails of ecology are intricately connected to the travails of women. The consequences of human disobedience mentioned in the narrative, which entail hostile weather conditions, violence between fellow human beings, violence among animal species, and general ecological degradation, are reminiscent of the current global ecological crisis and the ensuing socioeconomic inequalities, which have a huge bearing on women's welfare.

Below, we turn to discuss another Shona Indigenous creation narrative, the Guruuswa creation myth.

■ The Guruuswa Creation Myth

As noted by Cynthia Marangwanda,[14] the term Guruuswa refers to "long grass" or "tall grass." It is described as an area of grassy plains and expansive grassland. Most present-day Shona people trace their "historical" birthplace back to Guruuswa. David Mark Lan has the following to say about Guruuswa:

> In the study of social anthropology and Shona mythology, Guruuswa, the place of long grass, is both an existing geographical landscape in the rich grasslands of the Dande escapement, the habitant of Kwekwe and a symbolic fountain of fertility and resourceful motherhood. It is the figurative spring of life, the omnipotent wetland valley from whence issues all ancestry, all men and plant life.[15]

---

13. See Oduyoye, *Who will Roll the Stone Away?*; Amoah, "Theology from the Perspective of African Women"; Phiri, *Women, Presbyterianism and Patriarchy*; and Dube, *Other Ways of Reading*.

14. See Marangwanda, "Guruuswa."

15. Lan, "Making History," 152.

The above citation indicates that Guruuswa is to be conceptualized as both a geographical location and a spiritual space. For instance, on the one hand, Chirevo Kwenda describes Guruuswa as a physical space in the grasslands of the African Great Lakes region (East-Central Africa), while on the other hand, she identifies it as a mythical space, humanity's place of origin.[16] Hence, the term Guruuswa denotes the concept of origins in both an actual and a metaphoric sense, a real and symbolic way. In this light, David Mark Lan tenders that:

> The tall grass could suggest the symbolic pubic hair and the watery place indicating the female reproductive organ—the vagina [...] Guruuswa, then, is the source of social and biological life, a source of fertility within the lineage of the protagonist, Mutota. This implies the notion of motherhood which is consistent with the interpretation of Guruuswa as a vagina.[17]

Whether a mythical space or a physical location, what is apparent among the Shona is that Guruuswa denotes a pilgrimage from nonexistence to being alive for all life forms. For the Shona people, retracing their steps back to Guruuswa entails a spiritual journey. David Mark Lan summarizes the numerous versions of the Guruuswa creation myth as follows:

> The myths seem to say that the first man (the origin of social life) and the waters of the river (the basis of biological life) both travel towards the present home of the community from the source of the river where water rises out of the earth in a patch of long grass.[18]

Like the Mwedzi/Dzivaguru myth of creation, the Guruuswa creation myth also reveals a gendered disequilibrium. The main characters described in narrating the Guruuswa creation myth are either male or related through blood to Mutota, the male great ancestor. This point is reiterated by David Mark Lan:

> In no version of the myth is any reference made to a wife of Mutota or to any other affine. The people at Guruuswa and those who make the journey to Dande are agnates—Mutota's sons, his daughter and his brother, twin or friend Chingoo. The notion of twinship or of brotherhood emphasises the consanguinity that binds the other characters. In narrative terms, it provides companionship without introducing affinity.[19]

As noted in the above citation, the Guruuswa creation myth tends to peripheralize women's role in the whole drama of creation. Mutota is portrayed as multi-gifted; he is also depicted as a spiritual figure, one who does not need to or one who is destined to abstain from getting married.

---

16. See Kwenda, *The Guruuswa Myth of Migration*.

17. Lan, "Making History," 151.

18. Lan, "Making History," 152.

19. Lan, "Making History," 152.

Although the concept of Guruuswa is generally interpreted as a metaphorical depiction of the female reproductive organ, which is a source of conception and birthing life, the narrative of the myth skates over the feminine contribution to the emergence of life on Earth. This peripheralization of women is a recurring predisposition, not only within the Shona communal interactions but in most African Indigenous communities, where women are often pushed to the periphery of society. This marginalization of women and the patriarchal dominance over the earth are twin connections which need urgent redressing.

Below, we turn to a discussion of the various manifestations of gender and social inequalities within the Shona society.

■ Interrogating Inherent Gender and Social Inequalities Within Shona Society

The foregoing section has illustrated the entrenched patriarchal worldview exhibited in popular Shona creation myths. Cognizant of the fact that myths play a pivotal role in shaping people's norms and values, it follows that the gender and social inequalities embedded within these creation myths have a huge bearing on how society is structured. In the Zimbabwean context, two arenas where these gender and social inequalities are glaringly apparent are the traditional laws on land ownership and the misappropriation of culture to justify the gender-disproportionate distribution of natural resources such as mining rights.[20] For instance, within the Indigenous Shona community, the right to own land is passed from one generation to the other through male hierarchical structures. Hence, the son inherits land from the father, whereas the daughter cannot inherit her father's land. Whereas the Zimbabwean Constitution (2013) makes concessions for daughters to inherit not only land but other property from the father, within conservative Indigenous communities, particularly in rural areas, such a practice is unthinkable.[21]

The theme of male dominance has come out clearly in all three creation myths discussed (Mwedzi, Dzivaguru, and Guruuswa). The male characters in these creation myths are the ones who are elevated, and they are celebrated for bringing forth life. They are the ones who are given instructions by Mwari regarding the importance of ecological stewardship. Women's roles are either not mentioned (like in the Guruuswa narrative) or they are mentioned but not given the same prominence as their male counterparts (as is the case in the Mwedzi/Dzivaguru narratives).

---

20. See Chireshe, "Access to Land Ownership and Gender," and Chirongoma "Karanga-Shona Rural Women's Agency."

21. See Chireshe, "Access to Land Ownership and Gender."

This tends to be replicated in our contemporary times among the Shona people. For instance, whereas women do most of the backbreaking domestic agricultural work to ensure that their families and communities are well fed when it comes to land ownership, they get the shorter end of the stick. Women are also in the vanguard of ecological preservation—they are mainly responsible for tree planting, gardening, fruit farming, fetching water, cooking and serving food, collecting firewood, and caring for Mother Earth. When the land sickens, making people vulnerable to all manner of ailments, such as the coronavirus disease 2019 (COVID-19) pandemic, it is women who care for the sick, comfort the bereaved, and find ways to restore health and well-being, using natural remedies and other Western biomedical means.[22]

The apparent gender and social inequalities have far-reaching effects on the ecological crisis. We turn to deliberate upon that below.

■ The Encroaching Ecological Crisis in Zimbabwe

The menacing global ecological crisis has slowly but surely encroached on Zimbabwe's doorstep. Since the year 2000, Zimbabwe has been hit by several disastrous cyclones: Eline (2000), Japhet (2003), Dineo (2017), Idai (2019), Ana (2022), and the most recent, Cyclone Freddy (2023).[23] There have also been some disastrous tropical storms: Chalane (2020) and Eloise (2021).[24] These natural disasters have caused immense damage, not only to biodiversity but also the health and well-being of countless Zimbabweans. These natural disasters have a propensity for increasing people's vulnerability to epidemics and pandemics, especially those who already experience socioeconomic constraints.[25] The multifarious effects of natural disasters on people's health and well-being in Zimbabwe have been succinctly noted by Blessing Mucherera and Emmanuel Mavhura:

> Amongst all observed natural and anthropogenic adversities, floods and drought are the most recurrent hazards in Africa [...] Whilst changes in the environment are a source of exposure, sensitivity to these is the basis for defining the degree to which specific places are more or less vulnerable than others [...] Floods increase the vulnerability of rural households that mainly depend on rain-fed agriculture and livestock production [...] The most affected groups of people include women, children and the elderly.[26]

---

22. See Zvingowanisei, Chirongoma, and Chitando, "*Mbuya Nyamukuta* (Traditional Midwives)"; Zvingowanisei and Chirongoma, "Karanga Women and Indigenous Knowledge Systems (IKS)."

23. See Mukwenha et al., "Health Emergency and Disaster Risk Management."

24. See Chirongoma et al., "Reigniting the Principle of Ubuntu/Unhu."

25. See Mucherera and Mavhura, "Flood Survivors' Perspectives."

26. Mucherera and Mavhura, "Flood Survivors' Perspectives."

This citation affirms that women, children, and the elderly are more susceptible to the effects of ecological degradation. This calls for some urgent interventions to redress the harmful effects of the ecological crisis. Mothers are the nurturers and preservers of life. Children are the future, and the elderly are the doyens of wisdom. Hence, failure to curb the harmful effects of the pending ecological crisis is tantamount to failure to preserve the safety and security of the next generation.

Acknowledging the nexus of the ecological crisis and the outbreak of epidemics in Zimbabwe, Solomon Mukwenha and others propose the following contingent measures:

> Firstly, we recommend proactive immunizations against cholera, typhoid, and measles especially in the cyclone and tropical storm-prone regions. Secondly, in light of the current COVID-19 pandemic, it is also imperative to make provisions that allow for prompt resettlement of people in high-risk areas. The resettlement structures should allow for adherence to COVID-19 guidelines to prevent outbreaks. Thirdly, we call on the government to strengthen a multi-sectoral approach that includes the private sector emergency response in disaster preparedness efforts. Further, national budget reassignments coupled with improved health service provision methods can play a critical role in disaster preparedness and management. Lastly, the government of Zimbabwe should provide cyclone and storm emergency kits to individuals in cyclone-prone areas and also support the construction of houses that are resilient to climate disasters.[27]

The three recommendations in the above citation aptly speak to the realities currently pertaining on the ground. The ensuing ecological crisis requires a multisectoral approach. Making concerted efforts to address the ecological crisis, in tandem with a drive to enhance access to primary health care, particularly preventive health, will go a long way in reducing the gendered impact of the ecological crisis in Zimbabwe.

Besides the natural disasters, some of the losses suffered by the people of Zimbabwe in the past few years have been human-induced, mainly because of the unfair and unequal distribution of resources and opportunities. One of the controversial and disruptive government-induced exercises was Operation Murambatsvina (Restore Order), which was initiated by the government of Zimbabwe in mid-2005. This took place during the heart of a cold, windy, and rainy winter when the city council officials went on a rampage to destroy informal settlements and informal business structures. This operation not only destroyed poor and vulnerable people's homes, business premises, and their meager livelihoods, but it also stripped them of their sense of place. Suddenly, the affected people woke up homeless and bereft of their source of livelihood. The operation

---

27. Mukwenha et al., "Health Emergency and Disaster Risk Management," 1.

also entrenched the unequal distribution of resources, because most of the people who were affected by this exercise were already poor and under-resourced. Undertaking the exercise during the middle of winter also exposed the affected people to the vagaries of weather, exposing some to long-term and life-diminishing ailments.[28]

Another controversial exercise was the construction of the Tokwe Mukosi Dam, which caused the displacement of the Karanga people, who were forcibly removed to make room for the dam.[29] While the construction of this mega-dam will benefit the nation at large, it is sad that the process caused the Indigenous Karanga people to be uprooted and disconnected from their ancestral lands, which was their bastion of security. This caused them undue agony. I captured the experiences of the displaced communities elsewhere as follows:

> In the Karanga people's worldview, any misfortune and mishap is interpreted as a direct result of the separation between the living and their spirit elders. Relocating from one's ancestral land may be interpreted as abandoning the ancestors [...] Sacred places used by the traditional leadership such as sacred mountains were submerged in water, leaving the community in limbo as the meeting places with their spiritual powers disappeared. This has far-reaching repercussions on their spirituality since such a separation clogs lines of communication between the physical and the spiritual world. Furthermore, the community is thrown off balance as they are left without a claim to their space.[30]

What was even more heart-wrenching for the displaced community was that they did not directly enjoy the financial benefits wrought by the construction of the dam. This magnified their losses and entrenched the socioeconomic inequalities in society. Hence, being denied their sense of place, being forced to sever ties with their ancestral lands, and having to endure the loss of their livestock, household possessions, and agricultural produce during the relocation process was a bitter pill to swallow. Let me draw from my earlier work again:

> The wreckage wrought upon the Karanga people who were displaced to make way for the government initiated Tokwe Mukosi Dam, while not directly benefitting from the project, raises a clarion call to African Theologians to pay particular attention to protecting and preserving Mother Earth.[31]

The experiences of the survivors of Operation Murambatsvina and those who endured the Tokwe Mukosi displacements are only two examples of various incidents of displacement which caused communal disruptions and

---

28. See Chirongoma, "Operation Murambatsvina."
29. See Chirongoma, "Where Earth and Water Meet," 144.
30. Chirongoma, "Where Earth and Water Meet," 145.
31. Chirongoma, "Voices from the Margins," 166.

robbed Indigenous communities of their sense of place. Being uprooted from all that is familiar robs the affected population of their sense of belonging. It disrupts the place of their story and their story of place because among the Shona, one's identity and history are tied to a particular place, their ancestral home, where their *rukuvhute* [umbilical cord] lies buried.

In light of the foregoing discussion, the next section will propose a theology of responsible ecological stewardship as a response to the ecological crisis.

■ The Interface between Creation Theology and a Theology of Responsible Ecological Stewardship

As has been noted in the preceding sections, it is our responsibility as humans to serve as responsible stewards of Mother Earth. Failing to preserve Mother Earth, which is our abode, is at our own peril. In the last 20 years, Zimbabwe has endured countless ecological catastrophes. One of the most severe was Cyclone Idai, which made landfall in March 2019. It left a trail of destruction, especially in the eastern part of Zimbabwe. A study conducted by Ezra Chitando and myself reveals that most of the survivors of Cyclone Idai in Chimanimani and Chipinge feel that the occurrence of the cyclone was because of a breakdown of the relationship between members of the community and their ancestors. An overarching question being asked by the survivors of Cyclone Idai is: "What did we do to our mountain?"[32]

> This question denotes the Indigenous people's perplexity as they grapple with coming to terms with the calamity in their midst. Ngangu Mountain[33] which once represented a place for enrichment, sustenance, and was an oasis for abundant living, unexpectedly turned into a poisoned chalice and became an agent of death and destruction. Reminiscent of Mother Earth which is usually connected with nurturance and preservation of all life forms, Ngangu Mountain which was looked upon by the local people as their oasis of livelihood and well-being suddenly turned into a weapon of mass destruction [...] Whilst acknowledging that some individuals broke taboos and traditional prohibitions in their interaction with Ngangu Mountain and the sacred phenomena inhabiting the mountain, the survivors cannot help but feel let down by their beloved mountain and their ancestral spirits.[34]

---

32. Chirongoma and Chitando, "What Did We Do to Our Mountain?"

33. This is the name of the mountain whence the Cyclone Idai carnage was spewed in the form of landslides and boulders which destroyed thousands of lives, homes, and a great deal of valuable property in Chimanimani and Chipinge.

34. Chirongoma and Chitando, "What Did We Do to Our Mountain?", 86.

The above citation indicates that the survivors of Cyclone Idai interpreted the occurrence of this ecological disaster in a twofold manner. Firstly, they felt that the cyclone struck them because nature and the ancestors were punishing them for failing to be responsible stewards of the environment. Secondly, some of the survivors felt that because they had a cordial relationship with Ngangu Mountain, the catastrophe happened as an unprecedented act of the mountain letting them down.[35] Reflecting on the first interpretation elicits a theology of responsible ecological stewardship. This coheres with Kalemba Mwambazambi, who suggests that:

> African theologians, and indeed all humanity, should individually and collectively make a permanent commitment to environmental protection [...] African theologians and concerned people need to take into account and observe ecological norms as part of the fulfilment of their world mission to protect their environment according to the ordinances of God.[36]

The theme of responsible ecological stewardship in the above citation is one of the pillars in ecofeminist theology. I resonate with fellow sojourners on the path of ecofeminist theology in propounding a theology of responsible ecological stewardship and a theology of fair and equitable distribution of resources. Ecofeminism is premised on the belief that women and nature are intricately connected.[37] Hence, when the environment suffers any form of harm or degradation, women bear the greater burden because of their inseparable connection with the land. Acknowledging the pivotal role of women in ecological preservation as well as the importance of drawing from the wells of African Indigenous knowledge, Lilian Cheelo Siwila offers the following insight:

> [*There is*] a need to appreciate the value of Indigenous knowledge, especially in today's society where theologians are continuously challenged to seek other ways of practising theology [...] If religion is to meet the demands of its adherents, especially those at the peripheral of society, there is a need to embrace their value systems. Indigenous knowledge in this case is one value system that has been challenged by modernity and colonialism to the extent that some religions have threatened to do away with some of the practices that carry along valuable Indigenous knowledge.[38]

As has been noted in the first part of this essay, the Shona people's myths of creation offer important insights into their ecological perceptions

---

35. This echoes the same questions raised in this volume by Arnfríður Guðmundsdóttir in her essay where the people of Iceland grapple with the volcanic eruptions spewing from the mountains, which they once cherished as their source of sustenance and nourishment.

36. Mwambazambi, "A Theological View of Environmental Protection in Africa," 849.

37. This resonates with the essay in this volume by Melanie Harris, who reflects on ecowomanism as a pivotal resource for addressing ecological injustices.

38. Siwila, "Tracing the Ecological Footprints of our Foremothers," 137.

and interactions. This resonates with the above citation and the need for African ecofeminist theologians to pay particular attention to Indigenous Knowledge Systems.

Reflecting on the Cyclone Idai carnage, Ezra Chitando and I highlighted the following pertinent insights:

> From an African eco-feminist perspective, it is also important to interrogate the fact that women in the Global South are the ones who are bearing the brunt of the global ecological crisis. Such a status quo should spur African governments and African religious leaders to make concerted efforts towards putting measures in place to reduce the carbon footprint whilst paying particular attention to the plight of women and girls whose quality of life has been further diminished by the perpetually increasing ecological disasters.[39]

The above excerpt reminds us of the need to constantly interrogate global socioeconomic and gender inequalities, particularly how the ecological crisis places a heavier burden on women, especially those who are poor and reside in the Global South. A theology of responsible ecological stewardship, couched in fair and equitable distribution of resources, elicits the vision of *shalom*, enshrined in peace, justice, and equality. Emphasizing the importance of responsible ecological stewardship in Africa, Kalemba Mwambazambi offers the following propositions:

> African people need to apply ethical scientific knowledge for food production, to care for their environment by refusing to be polluting agents themselves, to live in a sustainable manner in accordance with the will of God and to avert the exploitation of Africa by foreign nations. Africans have to learn how to look after themselves in harmony with their neighbours. God has declared all creation good. Therefore, human beings should treat every part of creation as having its own intrinsic dignity. Africa still has time to choose for clean and renewable energy sources, and as a continent, should take the lead in an energy revolution designed to reduce climate change.[40]

This essay reiterates the need for a theology of responsible ecological stewardship as well as practicing fair and equitable distribution of natural resources. This is informed by the limited access to resources and opportunities given to the poor and the underprivileged, especially women and girls. This renders them vulnerable to the impact of ecological destruction. This point is reinforced by Ezra Chitando and myself:

> It is becoming crystal clear that climate change's most devastating effects will intensify among the world's poorest and most vulnerable communities. While those who have contributed the most to climate change, such as the rich communities of the Global North, experience the least impact, communities

---

39. Chirongoma and Chitando, "What Did We Do to Our Mountain?", 86.

40. Mwambazambi, "A Theological View of Environmental Protection," 851.

in the Global South will have to intensify efforts for survival. Therefore, it is important for activists at different levels to remind, particularly religious leaders, of their prophetic role.[41]

Ecological catastrophes have led people to raise theological questions regarding the belief in God's omnipotence and omnibenevolence. As noted by Chirongoma and Chitando, this is one of the recurring questions which was raised by the survivors of Cyclone Idai:

> Even those who uphold the Christian beliefs and values are left asking, "Where was God when this tragedy hit us?"[42]

Whenever disaster strikes, most African people turn to religion for answers. Hence, people will be left wondering, "Where was God when this disaster struck?" The question regarding the existence of evil and the goodness, power, and love of God is an age-old question. The Shona people in Zimbabwe are also among those who continue to grapple with such questions in the wake of ecological disasters such as Cyclones Eline, Dineo, Idai, or Freddy. This question needs constant reflection on our role and our place as responsible ecological stewards and the occurrence of some natural disasters which are beyond our control and surpass our understanding.

These questions dovetail with the same concerns raised within the field of creation theology. Reflecting on the nexus of *Creation, God, and the Coronavirus*, Mark Brett and Jason Goroncy[43] remind us that:

> Our planet continues to burn, and our seas continue to acidify. Patterns of production and consumption continue to despoil the resources needed by future generations. A new world is calling, for which we need a fuller understanding of creation.[44]

In concurrence with the above excerpt, it is pertinent for us to exercise our responsible ecological stewardship by revisiting God's original intent for humanity to "be fruitful and multiply and replenish the earth" (Gen 1:28). This entails embracing the fact that God's created universe must be protected and preserved. It reminds us that our place within the story of the universe is to serve as responsible gardeners, to reduce pain and suffering for both human and nonhuman entities. It also invites us to echo

---

41. Chirongoma and Chitando, "What Did We Do to Our Mountain?", 85–86.
42. Chirongoma and Chitando, "What Did We Do to Our Mountain?", 87.
43. Brett and Goroncy, "Creation, God, and the coronavirus."
44. Brett and Goroncy, "Creation, God, and the coronavirus," 351.

Ernest Simmons,[45] who reminds us that God is always present and shares in the travails of Mother Earth. He puts it across as follows:

> God has been entangled with a suffering creation from the very beginning, working from within the evolutionary process [...] It is here that the understanding of deep incarnation brings with it the intimate involvement of God through Christ in the very living cells of human flesh and all biological existence.[46]

Hence, in acknowledging God's active involvement in his created universe, we also acknowledge Christ's redemptive power in healing and restoring the broken universe and resurrecting fallen humanity through his death, resurrection, and ascension.

◼ Conclusion

This essay highlighted that creation myths help us to answer some of the fundamental questions of existence. The two overarching questions which informed the discussion are:

1. How does the Shona people's cosmology help us to conceptualize planetary, human, and cultural evolution?
2. How does the Shona people's ecological and cosmological purview help us to conceptualize the creative work of God's love?

In responding to these two questions, the essay explored the Indigenous Shona people's Mwedzi/Dzivaguru and Guruuswa creation myths. Drawing insights from these two creation myths, the second section of the essay interrogated the inherent gender and social inequalities in Zimbabwe, with a specific focus on land ownership and the distribution of natural resources. The third section discussed how the recurring droughts, cyclones, and erratic weather patterns in Zimbabwe are a manifestation of the global ecological crisis. On this basis, this essay offered insights on the need to constantly affirm and act upon the wisdom enshrined within creation theology, as well as a theology of responsible ecological stewardship. This coheres with Ernst M. Conradie, who tenders that:

> A theology of stewardship [...] suggests a harmonious and environmentally sensitive relationship between humans and creation. Human beings should be regarded as stewards, caretakers, priests, custodians, or guardians of creation [...] This fosters an environmental ethos where emphasis is placed on using resources wisely, sound management, reliability, commitment, dedication, hard work, and responsibility towards God the owner of the land.[47]

---

45. Simmons, "The Entangled Pandemic: Deep Incarnation in Creation," 2021.

46. Simmons, "The Entangled Pandemic," 356.

47. Conradie, *Christianity and Earthkeeping*, 81.

It is important to reiterate that the question of our "place" in the universe is a story of journeying and communing with Mother Earth; hence, failure to exercise responsible ecological stewardship may have catastrophic consequences, such as those that have besieged my motherland Zimbabwe. Reflecting on the incessant droughts, floods, cyclones, and mudslides causing a trail of destruction, I cannot help but join the other cloud of witnesses in questioning God's creative work and abundant love. If God created the universe and pronounced it as not only "good," but "very good," the searing questions that one continues to grapple with are, "What is happening to this 'very good' creation? Where is God when all these ecological disasters are happening?" Another closely related question is, "What is our place as humans in the created universe?"

My humble submission is that our place in the created universe is to care for and nurture Mother Earth. This is our divine commission. God's creative work is a constant reminder of our being inextricably connected to Mother Earth; it remains a "very good" gift bequeathed to us. This echoes Psalms 24:1: "The earth is the Lord's and everything in it." Hence, even when natural disasters happen, this does not dismiss the goodness and lovingkindness of God, the creator of the universe, who has invited us as humans to participate in communion with God's handiwork.

■ Bibliography

Amoah, Elizabeth. "Theology from the Perspective of African Women." In *Women's Vision: Theological Reflection, Celebration, Action*. Edited by Ofelia Ortega, 1–7. Geneva: WCC, 1995.

Brett, Mark, and Jason Goroncy. "Creation, God, and the Coronavirus." *Theology* 123:5 (2020), 346–52.

Chireshe, Excellent. "Access to Land Ownership and Gender in the Light of African Indigenous Religion in Zimbabwe amongst the Shona in Chiredzi District, Masvingo Province, Zimbabwe." In *Mother Earth, Mother Africa: World Religions and Environmental Imagination*. Edited by Sophia Chirongoma and Scholar W. Kiilu, 155–67. Stellenbosch: African Sun, 2022.

Chirongoma, Sophia, and Ezra Chitando, "'What Did We Do to Our Mountain?': African Eco-Feminist and Indigenous Responses to Cyclone Idai in Chimanimani and Chipinge Districts, Zimbabwe." *African Journal of Religion and Gender* 27:1 (2021), 65–90.

Chirongoma, Sophia, and Silindiwe Zvingowanisei. "Patriarchy as an Archetype of Empire among the Karanga People of Zimbabwe: Perspectives from African Women's Theology and Ubuntu/Unhu Ethic." In *I Can't Breathe the Knee of Babylon Is on My Neck: An African Perspective of Breaking Away from Babylon*, edited by Sindiso Jele, 35–60. Pretoria: Verity Publishers, 2021.

Chirongoma, Sophia, and Terence Mupangwa. "Gender Disparity as a Colonial Matrix of Power: Demystifying Pastors' Call Narratives in the Apostolic Faith Mission in Zimbabwe." In *Religion, Patriarchy and Empire: Festschrift in Honour of Mercy Amba Oduyoye*, edited by Lilian Siwila and Fundiswa Kobo, 211–60. Pietermaritzburg: Cluster, 2021.

Chirongoma, Sophia, et al. "Reigniting the Principle of Ubuntu/Unhu in the Aftermath of Cyclone Idai in Chimanimani, Zimbabwe in Light of the Sustainable Development Goals." *The Fountain—Journal of Interdisciplinary Studies* 3:1 (2019), 15–29.

Chirongoma, Sophia. "Karanga-Shona Rural Women's Agency in Dressing Mother Earth: A Contribution towards an Indigenous Eco-Feminist Theology." *Journal of Theology for Southern Africa* (Essays in Honour of Steve de Gruchy) 142 (2012), 120–44.

———. "Operation Murambatsvina (Operation Restore Order): Its Impact and Implications in the Era of HIV and AIDS in Contemporary Zimbabwe." In *Compassionate Circles: African Women, Theologians Facing HIV*, edited by Ezra Chitando and Nontando Hadebe, 71–94. Geneva: WCC, 2009.

———. "Voices from the Margins: Religio-Cultural Perspectives of Women, Children and the Elderly Amidst the Tokwe Mukosi Dam Displacements in Zimbabwe." In *Mother Earth, Postcolonial and Liberation Theologies*, edited by Sophia Chirongoma and Esther Mombo, 151–70. New York: Lexington, 2021.

———. "Where Earth and Water Meet: Exploring the Impact of Tokwe Mukosi Dam in Light of African Spirituality and Religion in Zimbabwe." In *Decolonizing Ecotheology: Indigenous and Subaltern Challenges*, edited by Lily Mendoza and George Zachariah, 146–61. Eugene: Wipf & Stock, 2022.

Conradie, Ernst M. *Christianity and Earthkeeping: In Search of an Inspiring Vision*. Stellenbosch: SUN, 2011.

Dube, Musa W., ed. *Other Ways of Reading: African Women and the Bible*. Atlanta: SBL, 2001.

Hargrove, Mavhu F. W. "Guest Voices: Shona Creation Story." Last accessed December 18, 2022. https://africancosmosdiary.wordpress.com/2012/05/18/guest-voices-shona-creation-story/.

Kalemba, Mwambazambi. "A Theological View of Environmental Protection in Africa." *In die Skriflig* 45:4 (2011), 849–66.

Kanyoro, Musimbi. "Engendered Communal Theology: African Women's Contribution to Theology in the Twenty-first Century." *Feminist Theology* 9:27 (2001), 36–56. https://doi.org/10.1177/096673500100002704

Kwenda, Chirevo V. *The Guruuswa Myth of Migration and the Quest for a Christian African Theology in Zimbabwe*. Indianapolis: Christian Theological Seminary, 1987.

Lan, David, Mark. "Making History: Spirit Mediums and the Guerilla War in the Dande Area of Zimbabwe." PhD Thesis, London School of Economics and Political Science, 1983.

Mahohoma, Takesure. "Experiencing the Sacred." *Studia Historiae Ecclesiasticae* 46:1 (2020), 1–17. https://doi.org/10.25159/2412-4265/3363

Marangwanda, Cynthia. "Guruuswa: The Origin and the Genesis." Last accessed March 25, 2023. https://mamoyoshrine.wordpress.com/2018/01/15/guruuswa-the-origin-and-the-genesis/.

Mucherera, Blessing, and Emmanuel Mavhura. "Flood Survivors' Perspectives on Vulnerability Reduction to Floods in Mbire District, Zimbabwe." *Jàmbá: Journal of Disaster Risk Studies* 12:1 (2020), a663. https://doi.org/10.4102/jamba.v12i1.663

Mukwenha, Solomon, et al. "Health Emergency and Disaster Risk Management: A Case of Zimbabwe's Preparedness and Response to Cyclones and Tropical Storms: We Are Not There Yet!" *Public Health in Practice* 2 (2021), 100131. https://doi.org/10.1016/j.puhip.2021.100131

Mwandayi, Canisius. *The Eyes of The Shona: Dialogue with Shona Customs in the Quest for Authentic Inculturation*. Bamberg: University of Bamberg Press, 2011.

Oduyoye, Mercy Amba. *Introducing African Women's Theology*. Cleveland: Pilgrim, 2001.

———. *Who Will Roll the Stone Away? The Decade of the Churches in Solidarity with Women*. Geneva: WCC, 1990.

Phiri, Isabel Apawu. *Women, Presbyterianism and Patriarchy: Religious Experience of Chewa Women in Central Malawi*. Blantyre: CLAM, 1997.

Scheub, Harold. *A Dictionary of African Mythology: The Mythmaker as Storyteller*. Oxford: Oxford University Press, 2000.

Simmons, Ernest. "The Entangled Pandemic: Deep Incarnation in Creation." *Dialog* 60:4 (2021), 351–59. https://doi.org/10.1111/dial.12699

Siwila, Lilian Cheelo. "Tracing the Ecological Footprints of Our Foremothers: Towards an African Feminist Approach to Women's Connectedness with Nature." *Studia Historiae Ecclesiaticae* 40:2 (2014), 131–47.

Taringa, Nisbert. "How Environmental Is African Traditional Religion?" *Exchange* 352 (2006), 191–214. https://doi.org/10.1163/157254306776525672

Zvingowanisei, Silindiwe, and Sophia Chirongoma. "Karanga Women and Indigenous Knowledge Systems (IKS): Towards Enhancing Agricultural Production and Food Security in Zimbabwe." *African Thought: A Journal of Afro-centric Knowledge* 2:1 (2022), 202–59 (Special Issue on "African Women and Indigenous Knowledge Systems," edited by Sophie Chirongoma, Ezra Chitando, and Mazvita Machinga).

Zvingowanisei, Silindiwe, et al. "Mbuya Nyamukuta (Traditional Midwives) as Reservoirs of Indigenous Knowledge Systems in Maternal Healthcare: A Case Study of the Karanga in Masvingo, Zimbabwe." *African Thought: A Journal of Afro-centric Knowledge* 2:1 (2022), 49–77 (Special Issue on African Women and Indigenous Knowledge Systems, edited by Sophie Chirongoma, Ezra Chitando, and Mazvita Machinga).

Reforming Place or Placing Reform? One Western Cape Version of the Story

Ernst M. Conradie[1]

■ Situating Creation Theology

All Christian theology is biography. This is widely affirmed in multiple strands of narrative theology,[2] whether evangelical, liberation, feminist, or decolonial in orientation—even though such a statement seems self-contradictory in claiming general validity. If theology is biography, it is also genealogy, situating one's story in a family history. For Christians, that family story goes back all the way to being adopted as children of Abraham and Sarah.

---

1. Ernst M. Conradie is a senior professor in the Department of Religion and Theology at the University of the Western Cape in South Africa.

2. For a discussion on narrative theology, see the introductory essay to the first volume in this series, entitled *Taking a Deep Breath for the Story to Begin...*

> **How to cite:** Conradie, EM 2024, "Reforming Place or Placing Reform? One Western Cape Version of the Story", in EM Conradie & WJ Jennings (eds.), *The Place of Story and the Story of Place*, in An Earthed Faith: Telling the Story amid the "Anthropocene", vol. 3, AOSIS Books, Cape Town, pp. 79-100. https://doi.org/10.4102/aosis.2024.BK355.05

All Christian theology is, in the same sense, also topography. It is situated discourse, assuming a sense of place, tied to a specific geographic location from which it may gain plausibility and credibility. In a fallen world, being located necessarily implies contestations over space. Place for some may mean displacement for others. Space thus implies a "site of struggle" involving issues of class, race, and gender. Such a sense of place should warn any theologian not to assume the universal significance of her or his story. Jerusalem, Rome, Constantinople, Alexandria, Carthage, Wittenberg, Geneva, Oxford, and Princeton may each be the center of the cosmos, but they cannot speak on behalf of the whole world already because of the particular language that is spoken. All theology is necessarily contextual, and those with universal pretensions need to be properly provincialized.[3]

However, theology is, in a third sense, also cosmology. In speaking of *God* (better: to God), it cannot speak of one place only as no place can be fixated.[4] It can only speak from one place but then has to relate that place to other places, even *all* other places. Such a sense of cosmic width should warn any theologian against attempts to control, even to *own* God, as expressions such as "my God," "my town" or even "my church" may mislead one to think. The Bible bears ample witness to how the personal is tied to the cosmic in the very same breath.[5]

So Christian theology is genealogy, geography, and cosmology. This will shape any response to the core question raised in this volume, namely, "What difference does it make to the story of cosmic, planetary, human, and cultural evolution to re-describe this as the creative work of God's love?"

In what follows, I will seek to heed such pointers in order to offer *a* story of place in the hope that it may be juxtaposed with other such stories in this volume. I will place this story within the Reformed/Reforming tradition, as I inherited that through my training at Stellenbosch University as well as in the Western Cape, both in the narrower sense of the University of the Western Cape (UWC), where I have worked since 1993, and in the broader sense of the Western Cape Province, where almost all my immediate ancestors come from and where I have lived almost my whole life. I will

---

3. For a discussion on doing theology, see the introductory essay to the second volume in this series, entitled *How Would We Know What God Is Up To?*

4. Alfred North Whitehead understood this already in 1933: "Modern physics has abandoned the doctrine of Simple Location. The physical things which we term stars, planets, lumps of matter, molecules, electrons, protons, quanta of energy are each to be conceived as modifications of conditions within space-time, extending throughout its whole range. There is a focal region, which in common speech is where the thing is. But its influence streams away from it with finite velocity throughout the utmost recesses of space and time." Quoted by Joseph Sittler in *Evocations of Grace*, 152.

5. On the "rhetoric of cosmic extension," see Sittler, *Evocations of Grace*, 111.

explore the ambiguity in the title of this essay, namely the typical tendency in the Reformed tradition to reform (or transform) the place where it is located, thereby reforming itself while sometimes failing to do justice to a sense of place, and which therefore is in need of a reminder to properly place such reforms. On this basis, I will return to respond to the core question raised in this volume by offering a constructive thesis on creation theology.

■ The Places of My Story

All four of my grandparents grew up on farms. Their parents were landowners but, at times, desperately poor given the wars of 1899-1902, 1914-1918, and 1939-1945, the Great Depression and frequent droughts and pests.

Friedrich Conradi (1668-1729) came from Marburg in Germany, where his father was a pastor and where he had gymnasium training. He settled in South Africa in 1685—at the age of seventeen. My paternal grandfather, E.M. (Ernie) Conradie (1893-1970), grew up in Mierkraal, in the Strandveld between Bredasdorp and Cape Agulhas, the southern tip of Africa. His father purchased the farm in 1898, and it is still in the Conradie family, now owned by my cousin (also E.M. Conradie) and his son. Mierkraal is the true home (*ikhaya*) of my early youth, a place to which I still return from time to time—mainly to weep at my father's grave. Except for the farm, my grandfather had very little in terms of material possessions but nevertheless excelled in the joy of giving. He married Jacoba Margaretha Uys (1895-1969) from Patryskraal, between Bredasdorp and Swellendam. Her paternal ancestor, Cornelis Janz Uys (1671-1714), came from Leiden, the Netherlands, in 1704 and settled in the Bredasdorp area from early on. Her father, Cornelis Janse Uys (1862-1899), died when she was four, leaving a wife and seven children with barely enough hay to feed the horses for a day or so. Two of her sons nevertheless studied medicine in Edinburgh in later years, presumably with scholarships!

My maternal grandfather was Nicolaas Everhardus Nel (1904-1976). The Nels (originally Néel) were Huguenots from Rouen in Normandy who came to South Africa in 1688, having been given 80 guilders (roughly half a year's salary) by the Dutch East India Company to settle at the Cape. My grandfather grew up on the farm Brakrivier between Calvinia and Nieuwoudtville (the country's wildflower capital) in the arid Northern Cape. His great-great-grandfather was Jacobus Nel (1768-1840), better known as Koos Kommandant, given his role in wars against the Bushmen (or San people, the derogatory term used by the Khoi-Khoi), who were native to such areas. My grandfather on this side married Huibrecht Maria Elisabeth de Beer (1903-1988), who grew up with her aunt but was reunited as a teenager with her own family living in Soutkuil near Verlorenvlei ("lost marsh"), in a rather arid area lost along the west coast. I visited this farm for the first time only as an adult. The first De Beer to come to South Africa was Matthias Andreas, who arrived here as a soldier from Lübeck in 1698.

All of these families were mainly of Dutch, French, or German origin, arriving at the Cape in the late 1600s or early 1700s. However, there are at least nineteen individual enslaved women in my ancestry, alongside Eva or Krotoa, two enslaved men, one Chinese and a few others "of unknown origin" (read: harbor children). Several of them appear multiple times in my genealogical records. I am, therefore, the descendant mostly of landed slave owners but also of landless slaves. The irony is unbearable: without slavery, I would not have been.

Ownership of these four farms only became possible on the basis of settler colonialism at the Cape, the nomadic lifestyle of the Khoi-Khoi, the epidemics that decimated the Khoi population, mixed with the slave trade, followed by feudal labor and military expeditions against the Bushmen (San). Although these four farms were all purchased by a later generation, individual ownership only became possible on the basis of colonial conquest. My grandparents' relative privilege in being children of landowners (as compared to their "Colored" farm laborers) was therefore built upon the colonial conquest of such land, albeit some 200 years earlier. Each of them was classified as "White" under apartheid race classification, thus becoming beneficiaries of the system in multiple ways and with a long-lasting impact.

Notably, none of these aspects form part of any family narrative. Instead, very hard labor, an ongoing struggle for survival, and missionary pietism form the undercurrent of such family narratives. With the advent of apartheid in 1948, most of the Conradies supported the more liberal South African Party, while most of the Uyses, the Nels, and probably the De Beers supported the National Party. However, for each of these families, religion was far more important than politics. Most of the women became teachers, while my maternal grandfather (who was a member of the Afrikaner Broederbond) and my own father became pastors. While only one grandfather held a university degree, each of my uncles and aunts (with one exception) studied at a university, and three of the nine hold a PhD. I would say that if you put a farmer, a pastor, and a teacher together, you may well get an ecotheologian.

After my father's death in 1966 (aged 34, an evangelical missionary, philosopher, and apartheid critic all at once), my mother (with my two sisters, one born after my father's death) moved to Stellenbosch, where I have lived ever since. Stellenbosch is an exceptionally beautiful town with some whitewashed Cape Dutch architecture, oak lanes, a well-known university, towering mountains, vineyards all around, and a river running through it. Growing up, this town was, for me, the center of the cosmos, with the Jonkershoek nature reserve as its sacred heart. Alongside Jonkershoek, the picturesque beach town of Kleinmond (where my Nel grandparents lived after retirement) and the Cederberg mountains with their rugged sandstone spots of beauty and wonder became my beacons. However, Stellenbosch is probably also the most unequal town in South Africa and possibly in the world, not only because of extreme poverty but also because several of the richest South Africans live there. As I have often remarked, my teenage and young adult years came with the recognition that the rest of South Africa is not as "beautiful" as Stellenbosch (to put this mildly) and that the town itself was built on colonial conquest, feudal labor, and domination on the basis of race, class, and species, all held together by a form of "Reformed" Christianity.

The UWC, where I started working in 1993, is only 25 kilometers away, but it is also worlds apart. It is a historically Black university, founded on the basis of race classification. Its history is one of protest against the charter on which it was founded, trying to create a society that has never been from an unattractive past and an ever-challenging present. Its identity does not lie in the past or the present but in the future. *Becoming UWC* is the apt title of a volume on its history.[6] Its motto is therefore telling: *respice et prospice* (looking back and looking forward). It is a precious experiment in social construction, creating a new world, a multiracial, multireligious microcosmos that has never existed before. Criss-crossing between Stellenbosch and UWC serves as a constant reminder to me of two worlds that cannot be easily reconciled with each other. Being in between these two worlds provides a sometimes creative tension within which my theological reflections emerge.

■ The Reformed Stories of Place That Shaped Me

I am standing (or falling?) in the Reformed tradition of Dutch origin as received, institutionalized, and embodied in the South African context since the mid-1600s. As a postgraduate student, I resisted the narrowness of my teachers' mainly Dutch and German conversation partners. I was challenged by the South African Black, liberation and Kairos theologies that I studied. I completed a PhD on David Tracy in 1991 and then turned to ecotheology. In the 1990s, I welcomed the influence of diverse thinkers such as Thomas Berry, John Haught, Sallie McFague, Arthur Peacocke

---

6. See Lalu and Murray, *Becoming UWC*.

and Rosemary Radford Ruether, alongside Jürgen Moltmann, Wolfhart Pannenberg, Joseph Sittler, and Arnold van Ruler. I read Barth and Bonhoeffer but did not find the key to doing ecotheology that I was looking for there. Instead, I found it in the notion of re-creation [Dutch: *herschepping*], as expressed by Abraham Kuyper and especially by Herman Bavinck and further developed, albeit in opposite directions, by Oepke Noordmans and Arnold van Ruler. This came as an uneasy surprise to me, as these were the very scholars who influenced my teachers.

What attracted me to the notion of re-creation was that it recognizes God's work of creation [*creatio*] as ongoing, especially in response to a fallen world. As an eschatological concept, it affirmed at the same time God's sustained loyalty to creation [*creatura*], to what is temporal, material, earthly, and bodily. Re-creation is therefore a way of holding together God's work of creation, salvation, and consummation without fusing them.[7]

By itself, this is surely highly attractive. However, the South African reception of this very tradition proved to be disastrous. It was the Reformed tradition of Dutch origin (represented by the Classis Amsterdam until 1824) that legitimized settler colonialism in South Africa (e.g., in the prayers of officials). British imperialism and colonialism were based on military and economic superiority but followed the same underlying logic, albeit now in Anglican garb. Afrikaner civil religion provided a framework within which white supremacy and patriarchy, dubbed *blanke baasskap*, could be taken for granted.[8] Accordingly, the "flame" of European civilization had to be kept alive amid the "sea of barbarism" in "dark" Africa. This flame was kindled by the elite culture of slave owners, the power of gunpowder, and the Bible.

To make matters far, far worse, the core concepts of apartheid theology, as developed in the 1930s, were derived from the neo-Calvinism of especially Abraham Kuyper but also of Herman Bavinck and the philosopher Herman Dooyeweerd. Volumes have been written in this regard.[9] Three comments may suffice here.

---

7. As far as I know, the term re-creation was introduced by Herman Bavinck, who saw God's work of salvation as a restoration of God's work of creation. For an excellent discussion of Bavinck's position, see Veenhof's *Nature and Grace in Herman Bavinck*. This notion of re-creation was radicalized by Arnold van Ruler in the sense that he described eschatological consummation consistently as re-creation and not as elevation, divinization, or new creation [*nova creatio*]. This emphasis is found throughout Van Ruler's oeuvre but especially in his *Verzameld Werk Deel III*. For a detailed discussion of Bavinck and Van Ruler's positions, see my *Saving the Earth*. A translation of twenty-one essays from this volume has recently been published by Wipf and Stock as *This Earthly Life Matters*.

8. Many scholars have recognized the role played by neo-Calvinism, German romanticism, Warneck's missiology focusing on ethnic peoples, and Scottish, Dutch, and Moravian pietism. See especially Bosch, "The Roots and Fruits of Afrikaner Civil Religion."

9. In my assessment, the best analyses are written from within and thus in Afrikaans. See especially Kinghorn, *Die NG Kerk en Apartheid,* and Coetzee, *Die "Kritiese Stem."*

Firstly, the policy of apartheid, as promoted by the Dutch Reformed Church during numerous conferences since the 1930s, was based on pragmatic considerations. Its main aims were to resist British hegemony and to address the "poor white" problem from within. This required a solution to the so-called native affair. That "solution" came in the form of apartheid and subsequently the Bantustan policies of the Verwoerd government. One may say that apartheid was a particularly crude application and extension of the British colonial policy, based as it was on racial segregation. Such pragmatism was not overtly theological, but it was expressed in the 1935 mission policy of the (white) Dutch Reformed Church.

Secondly, given such expedient pragmatism, apartheid theology is best understood as a second-order legitimation of a position already adopted on nontheological grounds. This also required exegetical support, given an alignment with Reformed Orthodoxy and its emphasis on the authority of Scripture, but again, such exegesis served the purpose of secondary legitimation.[10]

Thirdly, the categories employed to offer such a legitimation were derived mainly from neo-Calvinist sources, with specific reference to Abraham Kuyper's notion of sphere sovereignty and Herman Dooyeweerd's idea of law [*wetsidee*]. Again, volumes have been written about the creative (mis)appropriation of such concepts. The intellectual foundations of apartheid theology were laid by the philosopher H.G. Stoker and by *volksvaders* such as Totius. In essence, it expanded the notion of the orders of creation to include race and ethnicity. Each ethnic group needs to exercise sovereignty in its own sphere (hence the need for Bantustans). Any attempt at forced integration should be regarded as imperialist, symbolized by the tower of Babel. God stands on the side of (racial) diversity and enforces that (if need be) in order to protect minorities (like the Afrikaners) against (British) imperialism! Ethnic groups therefore need to be kept apart, if necessary, by imposing "law and order," as this is enshrined not only in historical developments (God's providence) but in the very orders of creation. In separation, a later separate development lies nothing short of "our" salvation—so that apartheid also operated as a quasi-soteriology.[11] A failure to maintain such separation can only breed conflict. Because ethnic peoples and civilizations are so different, for the sake of peace, being apart needs to be protected—through law and, if necessary, through military means. This is a classic case of domination on the basis of the social construction of such differences. The earliest expression of apartheid as a theology may be found in the 1930s, but it

---

10. See also Loubser, *The Apartheid Bible*.

11. See Coetzee and Conradie, "Apartheid as Quasi-Soteriology."

culminated in *Ras, Volk, en Nasie* ("Race, People, Nation"), published in 1974, ironically when apartheid as the dominant political ideology was just about to be replaced by an emphasis on state security following the Soweto uprisings in 1976.

In South African Reformed theologies from the 1970s onwards, apartheid theology was vehemently resisted from within by scholars such as Allan Boesak, David Bosch, Nico Botha, Russel Botman, Jaap Durand, John de Gruchy, Shun Govender, Willie Jonker, Johann Kinghorn, Adrio König, Klippies Kritzinger, Chris Loff, Takatso Mofokeng, Beyers Naudé, L.R.L. Ntoane, Dirk Smit, Nico Smith, Amie van Wyk, and several others[12] (notably all male, but not all white). The sources of inspiration for many of these scholars came mainly from Karl Barth and Dietrich Bonhoeffer. This came to fruition in the declaration of a *status confessionis*, the dismissal of apartheid theology as a heresy, and the Confession of Belhar as formulated in 1982 and endorsed by the former Dutch Reformed Mission Church in 1986. This revolved around a unique critique of apartheid, namely that the political ideology assumed the fundamental irreconcilability of people in terms of race. Accordingly, people are so different that it is best, for the sake of peace (!), to keep them apart. Once this ideology is affirmed or legitimized in the church, it is directly in conflict with the message of reconciliation in Jesus Christ, with reconciliation in the church, and with the ministry of reconciliation in society. This critique was first articulated in class discussions at UWC in 1978. Lecturers at the UWC (bordering on the suburb of Belhar) were deeply involved. As a Stellenbosch student and later as a lecturer at UWC, I wholeheartedly endorsed the Confession of Belhar.[13]

The Confession of Belhar, with its distinct emphasis on ecclesial unity, reconciliation in Christ, and a God of justice, provided me with tools to extend such a notion of justice toward ecojustice (responding to both economic inequality and environmental destruction). It also allows for a critique of consumerist culture.[14] However, there is no reference to creation in the Belhar Confession, and it therefore cannot (and need not) suffice for a more fully developed form of ecotheology. I found Barth's critique of natural theology (in the Germany of the 1930s) legitimate but not his alternative in seeing creation as the external basis of the covenant.[15] Instead,

---

12. While other names could readily be added, I did not include the figure of Johan Heyns here because his position was ambiguous and open to interpretation. For a discussion, see Conradie, "The Ambiguity of Johan Heyns."

13. For an authoritative discussion, see especially Cloete and Smit, *A Moment of Truth*.

14. See my *Christianity and a Critique of Consumerism*.

15. For a discussion, see the volume that I edited entitled *Creation and Salvation: Dialogue on Abraham Kuyper's Legacy for Contemporary Ecotheology*. Barth himself was ill at ease with theme of creation, as he admits in October 1945 in the preface to Volume 3.1 of the *Church Dogmatics*, ix–x.

I found myself returning to Bavinck's notion of re-creation, as developed in his *Reformed Dogmatics* and in *Philosophy of Revelation*.

This left me with me an intriguing set of questions. How could a theology that saw itself as staunchly (and rigidly) Orthodox become so completely heretical? How could the very same sources be used to legitimize apartheid *and* prove liberative in the struggle against it? Should one simply abandon neo-Calvinism as hopelessly contaminated? Could one perhaps argue that Kuyper's theology was misunderstood? But why was he used in the first place in support of racist politics? Is there not some contamination, guilt by association? Are opposing readings of Bavinck's *Reformed Dogmatics* really possible and plausible, that is, by sitting to Bavinck's left or to his right? Would going back to Calvin, and before him to Augustine, perhaps help to resolve such questions?

It became clear to me that there are very different views (stories) of creation (place) within the Reformed tradition of Swiss, Dutch, and German origin and its reception in the South African context. Likewise, there are radically different interpretations of the notion of "re-creation." Upon this recognition, I wrote *Saving the Earth?* (2013), in which I analyzed the diverging views of Calvin, Bavinck, Barth, Noordmans, Van Ruler, and Moltmann on re-creation. The volume concluded with a chapter on the reception of this tradition in South Africa and an assessment of what my South African teachers handed to me and my generation in this regard.

These, then, are the Reformed stories of place that continue to influence my engagements with ecotheology in general and creation theology in particular. Clearly, some interplay is needed between attempts at reforming place and placing reform, but this interplay can easily go wrong. How, then, should one proceed? What would help to avoid further distortions? With the thesis on creation theology that I offer, I am seeking a way forward to address this disturbing question. There can be no guarantees, but the typically Reformed dynamic of Word and Spirit, questioning any rigid form of Orthodoxy, together with mutually critical ecumenical engagements (as in this volume), would surely help.

■ Storying Place

Any creation theology requires a distinction between Creator and creation, as well as a distinction between creation as act [*creatio*] and as 'fact' [*creatura*] (see the similar distinction between act and being). The latter distinction coincides in this volume with the interplay between story (act) and place (outcome). For creation as act, a further distinction is often made between creation in the beginning (e.g., *ex nihilo*), ongoing creation/becoming/evolution/process [*creatio continua*] and consummation (eschatological re-creation)—if you like, the past, present, and future dimensions of creating.

Each of these distinctions has led to long-standing debates, also in ecotheology. Such debates cannot be resolved or even addressed here. Suffice it to say that such distinctions remain necessary in order to confront the confusion so typically associated with creation theologies. In earlier work, I suggested that the question is what the actual question is that is addressed in creation theology.[16] Some focus on the question of whether the world was created, typically in conversation with philosophies from Plotinus to Spinoza to Whitehead. Some focus on *how* the world was created, typically in conversation with the natural sciences from Galileo to Darwin and beyond. Some, following Schleiermacher, focus instead on the question of what being a creature may mean in dependence on the Creator. Others, following Barth, resist such an emphasis on origination and dependence and insist that the doxological focus should be on *who* created, on the identity and character of the Triune Creator.[17] On this basis, yet others speculate about the question why God created the world—for God's own glory, for "man" and his benefit, for the sake of love, or for the fun of it.[18] There are others who explore the meaning of creation and creativity in relation to becoming, procreating, birthing, mothering, evolving, ordering, developing, maturing, manufacturing, enabling, and so forth.[19] Then there is the more traditional focus on *what* God created: the visible and the invisible, Earth and heaven, plants, animals, humans, and angels, or life in general, and therefore time itself.

This focus on what God created is, of course, crucial for any ecotheology, but it harbors many dangers. One is the anthropocentric focus on humans as the "crown of creation." Another is idle speculation on the invisible (angels). A third is that theological content is emptied by drawing on and reiterating common human experience or scientific knowledge of the world around us—posing that as what God created and what that may mean. An example is talking about being the image of God without reference to God's identity and character.

---

16. See Conradie, "What on Earth Did God Create?"

17. This is neatly captured in the title of Colin Gunton's *The Triune Creator.*

18. The suggestion that God created the world for the fun of it is derived from Van Ruler. He often repeats that God created for the fun of it. We are because God cracked a joke, a good one. He would say that the Apostle Paul was correct in highlighting the importance of faith, hope, and love. But, he would add, he is sure that the apostle would agree upon further reflection to add joy—and would concur that joy is a higher virtue, even if compared with love. See Van Ruler's essay "Joy as the Quintessential Christian Experience of Life" (in *This Earthly Life Matters*) and Dirk van Keulen's excellent essay "*We zijn een grap van God*" ("We are God's joke").

19. For one erudite example, see Keller, *Face of the Deep.*

A fourth danger is particularly serious, namely that the focus on what God created can be used to legitimize the status quo of domination in the name of differences of race, gender, class, and species. From the time of Solomon to Constantinian Christianity (the Byzantine Empire) and the successive superpowers that emerged since the Renaissance (Italy, Portugal, Spain, the Netherlands, France, Britain, Germany, the United States of America, and to some extent Russia), there is a striking correlation between interest in the theme of creation and positions of power.[20] If such theologies of creation do not explicitly legitimize imperial and later colonial conquest, they often assume and reflect relative affluence. Put differently, the theological critique of such conquests typically does not come in the form of a creation theology. At worst, this yields a theology of blood and soil, apartheid theology, or a legitimation of an oppressive order based on class or caste, one where landlords see the beauty of God's creation while feudal workers and the landless unemployed only see gates and fences designed to keep them out.[21]

The same underlying logic is expressed in different contexts: God created the world as "we" know it. Despite the impact of sin, God maintains an underlying order to curtail its further rapid spreading. This order covers issues of gender (patriarchy), race (white supremacy), class (aristocracy), caste, and culture (hegemony). Resistance against such order is resistance against God as Creator. "We" are God's representatives on Earth, called to establish law and order, to "rule" over the land, for the sake of bringing about God's reign on Earth as it is presumed to be in heaven. As the whole earth belongs to God, "we," as God's representatives, are called to establish God's reign throughout the world. Because the land belongs to God, it also belongs to "us" as God's representatives.

In order to maintain a focus on what God created, while heeding such dangers, it is, in my view, necessary to story place, to tell the story of any one place. One may zoom in or out, telling the story of one individual in one particular location or even the universe story (as with Thomas Berry and his followers). Any such a story needs to account for how things are at the moment, how things were in the past, how it could be, and how it should be. There may be multiple lenses and multiple versions of the story, but these need to be connected by a sense of place, tied to a particular location. There may be his-stories and her-stories. (Evolutionary) science may play a role, as could the arts and humanities, culture, and religion. The characters in the story could include plants, animals, humans, angels, and demons. Stories may complement each other, but contestation should be expected and welcomed.

---

20. Likewise, geography and cartography emerged with imperialism. See Westhelle, *Eschatology and Space*, 17; also Jennings, "Reframing the World."

21. See again Westhelle, "Creation Motifs."

Christians may contribute to such stories of place, knowing that they do not have a monopoly over such stories, over their interpretation, or over the place itself. They have to do justice to how things are at this moment in time and then cannot impose their story/stories on others or on a particular place. They cannot assume the validity of their story/stories in advance but must follow the longer route of making sense (involving all the senses) of that place. Yet, it would not help to merely repeat the stories of others. What, then, can Christians contribute to the collaborative task of sense-making?

In an earlier collaborative research project on creation and salvation, entitled "The Earth in God's Economy" (2006–2015), I came to the conclusion that justice can only be done to both creation (as *creatio*) and salvation if these two aspects of God's work are held together by God's work of consummation and indeed by all the other aspects of God's work. These include seven "chapters" of the story, namely creation; ongoing creation (evolution); the emergence of humanity in all its grandeur and misery (the fall); God's providential care to curb the subsequent impact of evil (providence as *conservatio*, *gubernatio*, and *concursus*); God's election and the subsequent history of salvation; the establishment, formation, up-building, governance, ministries, and missions of the church with its sacraments; and the consummation of God's work. One may debate the terms used and the narrative sequence, but to fuse everything into one would spoil the story. The one aspect has to be balanced by the others. As with a juggler, it does not matter with which "cone" one starts, as long as all of them are kept in play.[22]

Three further moves followed from this conclusion.

Firstly, the classic notion of God's economy [*oikonomia tou theou*, i.e., the economic Trinity] places all these aspects within a narrative framework. Each of these aspects of God's work serves a heuristic function, namely as a tool that may be employed to help make sense of the world around us. Together, they provide an exceptionally rich set of narrative devices for the task of sense-making.

Secondly, although an arc from creation to consummation is evident, there is no master narrative and no need for that. There are multiple ways of telling the story, in different confessional traditions, Christian communities, extended families, and geographical contexts. Theologians each tell the story in their own ways, highlighting some features more than others.[23]

---

22. This is developed especially in my *The Earth in God's Economy*.

23. See the colloquium on "How Are They Telling the Story?" that followed from this insight, published in two volumes of the journal *Scriptura*.

There may be as many ways of messing up the story, so that it is not a matter of anything goes. The story has to be told in such a way that it does help those listening to make sense of their place.

Thirdly, the obvious needs to be made explicit: for Christians this remains a story of who *God* is and what this Triune God is doing at this point in time. It is a matter of discerning the signs of the times and the counter-movements of God's Spirit. Without such a *theo*logical focus, the story is reduced to running political commentary (or gossip) on unfolding events. Whatever version of the story is told by some Christians, this needs to be tested ecumenically with reference to God's identity and character. The clue to the story of the economic Trinity is related to the mystery of the immanent Trinity.

It should be evident by now that this series on "An Earthed Faith: Telling the Story amid the 'Anthropocene'" builds upon such insights, here framed as "storying place."

■ Placing Story: A Constructive Thesis

As suggested above, it would be inappropriate to take a theological shortcut to tell the story in an abstract and generalized way. It only helps to story place if one can place the story, that is, if the narrative framework can help to make sense of a particular place, the place in which one finds oneself. Creation theology cannot take a point of departure in what the world was presumably like "in the beginning" or in a utopian ("no place") vision of how it should be. It must focus on the world as we find it around us at this point in time. It must provide a sense of reorientation amid disorientation. It does not help to reconstruct how things presumably were, as none of us were there in the beginning, and the world around us has been evolving ever since. It is intellectually dishonest to focus only on what one finds attractive now (the beauty of nature) and to then declare that this must have been what God created. Instead, one has to take the world as it exists now seriously, in all its grandeur and its misery, its pain and its panache.

In a presentation prepared for the World Council of Churches' (WCC's) assembly in Busan (2013), I proposed a thesis on this basis that I have developed in various ways ever since.[24] For the sake of further ecumenical

---

24. This section builds upon my earlier work in "What on Earth Did God Create" and *The Earth in God's Economy*. See also my article on "The God of Life." More recently, this thesis was developed in a monograph in Afrikaans entitled *Kontra-intuïtief* ("Counter-intuitive") to indicate that the confession that *this* world is God's beloved creation is both counter-intuitive and profound.

discourse, this thesis may be formulated in the following way from within a Western Cape context:[25]

> The confession of faith in God as the Triune Creator offers a profound but deeply counter-intuitive liturgical redescription and ascription of the world, as we find it around us, at this point in time, seeing the world, prospectively and retrospectively, as it were through God's eyes.

Let me unpack the significance of this thesis with regard to each of the key terms, noting again my own situatedness as described above. This may enable me to see some aspects clearly but will also distort and hide aspects that would need to be corrected by others.

a) Creation theology is not only or even primarily about the ultimate origins of the universe. We do not know that, what, or how God created "in the beginning," as we were not there as witnesses. It concerns *the world as we find it in and around us*, in what is awesome and awful, ordinary and extraordinary, palpable and hidden from our eyes and ears, filled with joy but also with sorrows. It is inappropriate to focus only on what is beautiful, magnificent, and fruitful by claiming that this is what God created. The world as we find it around us includes pain and misery (not only for humans), poverty, injustice, and oppression, disability and deformation, anxiety and fear. The tension is perhaps most evident in the food cycle, where food for predators implies being eaten for their victims.[26]

b) To say that such a world is created by God is immediately and deeply *counterintuitive*. It may well sound more like an accusation than a form of praise. It seems to suggest that God is responsible for all of that—if perhaps not for injustice and oppression (for which humans need to take responsibility), then at least for inequality (and all that entails) and misery. Given the apparent implication of such inequality, namely that God has favorites, this invites the theodicy question. The immediate response should be, "How could God have created such a world?" Not just, "How could God allow suffering to prevail?" Because the confession is so counterintuitive, a distinction is necessary between what reality is and what it ought to be. This is appropriate, but to retroject what we now think ought to be into the distant past, thinking that this must have been what God created, is a social construct. This may be unavoidable, but the danger is that one may then eschew what is material, earthly, and bodily—as beloved by God.

---

25. The discussion of this thesis in this section builds upon more or less concurrent work published in Conradie, "Doing Creation Theology in the South African Context," with some amendments to fit the argument in both cases.

26. I have discussed the significance of predation in several essays, including "Could Eating Other Creatures Be a Way of Discovering Their Intrinsic Value?"

c) To affirm that this world is God's creation therefore immediately requires further explanation, *looking back and looking forward*,[27] namely by telling the rest of story, from the "very beginning," following subsequent trajectories leading to the present situation, always with an eye to what the future may hold. This is a story of grace and tragedy, providence [*pro-videre* = to see ahead], and a cloud of unknowing (not seeing), sin and redemption, the spread and curtailment of evil, a story about Israel, church, mission, and society. This story remains incomplete, unfinished, open-ended. Christians know even less how the story will end than how it all started. For these reasons, creation theology can never stand on its own. The theme of creation needs to be juggled with those of ongoing creation, human emergence, sin, providence, salvation, the formation of the church and its ministries and missions, governance, and the coming of God's reign, on earth as it is heaven. Again, for a juggler, all the cones need to be kept in the air. To catch the one is to let all the others fall.[28] The crucial question here is how to do justice to God's work of both creation and salvation, to the first and the second articles of the Christian creed. This is far easier said than done.[29] This raises further questions on how the story is to be told. Clearly, it can be told in many ways, over many Arabian nights, but it can be messed up in even more ways. Perhaps we do need to take a deep Breath for the story to begin...[30]

d) To see this world as God's own creation is to offer a *re-description* of the world. It is not only a matter of seeing but also of "seeing as." It is to see the world, as it were, through God's eyes. This is the purpose of the Christian liturgy, namely for believers to slowly turn their eyes away from dominant patterns of seeing the world in society in order to see the world in the light of the Light of the world. These dominant patterns suggest or assume that the world tacitly belongs to those who are strong, clever, rich, fast, dexterous or beautiful. To learn to see the world through God's eyes is therefore deeply counterintuitive, requiring a slow, patient process, a spiral of liturgical orientation, disorientation and reorientation.[31] This emphasis on "seeing as"[32] is distinct from natural theology in the sense that the latter looks at the world to find evidence for God's existence (but hardly God's character), while the former looks

---

27. The motto of the UWC is *"respice et prospice."*

28. For the image of the juggler, see again my *The Earth in God's Economy*.

29. For an exploration of the underlying problem, see the two volumes that I edited on *Creation and Salvation* (2011, 2012), and my essay "What Is the Place of the Earth in God's Economy?"

30. See again the title of the first volume in this series on *An Earthed Faith*.

31. I am drawing here indirectly on the work of Gordon Lathrop. See, for example, his *Holy Ground*.

32. On seeing nature as creation, see the argument of my *The Earth in God's Economy*.

at the world through God's eyes and character to recognize a beloved child.[33] As a theological category, it is also distinct from discourse on the social construction of reality (a sociological category) or from the role of worldviews (*Weltanschauungen*, a philosophical category) or cosmologies (a scientific category).[34] Seeing the world as God's creation must be realistic in the sense that it must *see* the world as it actually is (and not only how it should be or in the sense of wishful thinking). Yet to see *as* implies to see what is precious, to see possibility and promise. The emphasis remains on seeing the *world* and not on looking at one's *seeing*. The focus is on interpreting the world but then in the light of the word.[35] It requires what Jürgen Moltmann describes as a hermeneutics of nature (finding God's footprints and signature) but also the treacherous task of reading the "signs of the times," discerning God's hand in history.[36] This yields what may rightly be called a theology of nature as distinct from a natural theology.[37] As nature is intrinsically historical, this also yields a theology of evolution and indeed a theology of history. This is not merely an intratheological task: the focus must remain on interpreting nature. The question is whether theological reflection can add value to insights derived from other disciplines, in conversation with such disciplines.

e) The re-description of the world around us as embedded in the Christian confession is not only deeply counterintuitive; it is also absolutely *profound*. Its significance cannot be reduced to notions of origination and dependence although it includes that. Indeed, the world is not an autonomous entity that has its origin, life, and destiny in itself.[38] It belongs to God as its Creator; it is wanted and beloved by God. It is profound not merely because that which is beautiful, good, and true comes from God and belongs to God. It is expressed in the heart of the gospel, namely that God loves this world (as it is) so much that God's only Son came to save it. In short, in God's eyes, that which is material,

---

33. See Fensham, "The Sacrament of the First Child of God."

34. The term "worldview" is used in diverging theological discourses, and in South Africa this is often shaped by neo-Calvinism, where the distinction between a worldview and the doctrine of creation easily becomes conflated. See my reflections in "Views on Worldviews."

35. Herman Bavinck regards the task of a "philosophy of revelation" as an attempt by Christians to reflect on the significance of God's whole revelation, to relate the idea of revelation with the rest of our knowledge. See his thought-provoking and still underappreciated *The Philosophy of Revelation*. Suffice it to say that his vision remains elusive.

36. See Moltmann, *Sun of Righteousness, Arise!*, 189–208. He also employs the distinction between seeing the world as nature and as creation.

37. See my essay "Unravelling Some of the Theological Problems Underlying Discourse on 'Nature.'"

38. See the Faith and Order Study Document, *Confessing the One Faith*, 35.

bodily, and earthly is precious to the Father, is worth dying for to the Son,[39] and is being sanctified by the Spirit. The point is that this applies also to that which seems a lost cause—exterminated species, desolated lands, poisoned environments, people with severe disabilities, children born with deformities (consider fetal alcohol spectrum disorder), the proverbial second pelican chick, the victims of history, the tortured, irritating beggars, those addicted to crystal meth (*tik*), uncouth youth, you name it. How could these be God's creatures? How could God love them that much? It is such love that made Jesus of Nazareth a scandal to others. It may even apply to perpetrators, torturers, oppressors, rapists, and corrupt state officials. In words reminiscent of Desmond Tutu, it is to see the beggar as one's brother, the prostitute as one's sister, and the rapist as one's uncle. The harsh realities must be confronted, but how one perceives them makes a world of difference.

f) The re-description of the world as God's creation includes the connotation of *ascription*. The world is God's doing and belongs to God. This deepens the counterintuitive nature of the confession. If this is God's doing, how could the world end up being like this? How could it belong to God if it looks like it does? If those starving from hunger are indeed God's children, why does God not look after them as any capable parent would? If this is God's own garden, why is it in such a mess? Moreover, how could the world (as it is) be ascribed to *this* God, that is, given the identity and character of this triune God. One may perhaps expect such a world from a god who needs to be bribed to ensure fertility and favor. Or from the conflict between an array of unsavory divinities. Or if god was nothing but a proverbial "wicked old witch." But surely not this God?! The identity and character of this God therefore matters; it makes all the difference to whom the world is ascribed. If the world is indeed a self-autonomous entity that emerged merely by chance, then that has far-reaching implications for how one is to live in such a world. Perhaps one must, then, as it were, take a chance. If omnipotence and omniscience are ascribed to God, then this may well yield a world of power and surveillance where the wicked are punished mercilessly. The Jewish and Christian traditions only gradually and grudgingly came to the conclusion that this God is a God of love, peace, and joy, a God of mercy for the weak and vulnerable, and precisely for that reason, a God of justice who engages with perpetrators for the sake of their victims. If so, the polemic nature of such ascription should be recognized. Already in Genesis 1, the affirmation is that the world belongs to the God of Israel as its Creator and therefore not to the Babylonian Marduk, or the gods of any other empire. This polemic thrust may sound arrogant in the context of

---

39. See Hall, *The Cross in Our Context*, 24, 31.

interfaith dialogue and the need to respect the faiths of others. However, it should also not be domesticated. To ascribe the world to a God of mercy and justice is to deny that it belongs to the colonial gods of power, money, and Empire.

g) One may conclude that creation theology therefore has past, present, and future dimensions. It is about ultimate origins, but also about absolute dependence, about vocation (where creatures become co-creators)[40] and about ultimate destiny. Put differently, creation theology includes an emphasis on God's sustained faithfulness in suffering; the dangerous recollection of God's original intention for creatures; and the encouraging promise that the whole of creation will be liberated in Christ and through the Spirit.[41] This may, in my view, indeed be captured in the term "re-creation" (not to be confused with or reduced to recreation).[42] The hope is that God is renewing the whole of creation through the healing of distorted, broken relationships. These three temporal dimensions are held together in the present by the Sabbath as the true crown of creation, symbolized by the biblical seventh day.[43] This is the day when Creator and creature, the whole of creation, gather to enjoy one another's company; for humans (in some cultures) to eat bread and drink wine. This is the ability to enjoy such rest even if surrounded by suffering because of such a dangerous recollection and an encouraging promise, even though there has never ever been such a day of rest in the history of cosmic, biological and cultural evolution.

■ Seeing the World as God's Creation: The Difference That Makes a Difference[44]

What difference, then, does it make to the story of cosmic, planetary, human, and cultural evolution to re-describe this as the creative work of God's love? To see the world as the Triune God's beloved creation is very different from seeing nature in a number of other ways:[45]

---

40. The suggestion that creatures (not just human creatures) are co-creators with God is expressed by Michael Welker in *Creation and Reality*. Let the earth bring forth! See also Keller's *Face of the Deep* and Philip Hefner's more anthropocentric notion of the created co-creator in *The Human Factor*.

41. See the profound analysis by Gerard Liedke and Ulrich Duchrow in *Shalom*, 47–75.

42. See again my *Saving the Earth?*

43. See especially the argument in Moltmann's *God in Creation*.

44. For the concept "the world as God's creation," see also Volume 2 in the series "an Earthed Faith: Telling the Story amid the 'Anthropocene'," namely Conradie and Moe-Lobeda, *How Would We Know What God Is Up To?* (2022).

45. See the discussion of various "warped views" of nature as identified and described by Howard Snyder in *Salvation Means Creation Healed*, 42–45.

- The Gnostic, Manichean, or elitist view of nature as something essentially inferior to which value must be added, as something in need of being elevated, spiritualized, or divinized.
- The romanticized view of a leisured middle-class seeing nature as "oh so beautiful."
- A social Darwinian understanding of nature as "red in tooth and claw" and of society in terms of a bloody power struggle where the motto of the survival of the fittest reigns supreme.
- The colonial and capitalist notion that land is open to be conquered, that it is nothing but real estate, that nature provides "natural resources" that are available for excavation for the sake of economic development through scientific research and technological progress.
- The consumerist mentality that the world is there to provide us with pleasure and ecstatic experiences mediated by sophisticated forms of technology, seeing places as exotic tourist attractions, where there are preferably no beggars in sight.
- A pantheist vision of nature as something so sublime that it ought to be treated with reverence, if not worshipped.

Compared to such alternatives, seeing the world in terms of the Christian confession is highly attractive, also from an ecological perspective. The ecological significance of the Christian confession is best expressed through a number of models and images, each with biblical roots, with some considerable strengths if also some dangers.[46] Consider seeing the world as a fountain of life flowing forth from God; as God's fragile clay pot; as God's sacramental gift (to humans?); as God's work of art; as a musical composition, portraying God as an improviser with unsurpassed ingenuity, engaged in composing, choreographing, and performing the *opera trinitatis*; as God's theater; as God's playground; as God's beloved child; as God's own intimate body; as God's perichoretic dance partner; as God's bird's nest[47]; or as God's household [*oikos*].

God's household is perhaps the dominant image in ecumenical circles, thus linking inhabitation, political economy, and ecology through the common root of *oikos*. This way of seeing the world implies attachment rather than ownership, a sense of belonging, connectedness, and participation that addresses forms of alienation and exclusion. With Upolu Vaai, it suggests that "We are Earth"—alive, connected, and active. "We are the walking Earth and Earth is the ancestral us."[48] However, it then remains

---

46. I discussed these images in more detail in "What on Earth Did God Create?," with multiple references to the literature.

47. See Wallace, *When God Was a Bird*.

48. See Vaai, "We Are Earth," 72.

crucial to ask, with James Cone, "Whose Earth Is It Anyway?", and then with George Zachariah, "Whose Oikos Is It Anyway?"[49] The polemical confession that the household belongs to God may be profound, but only if the conviction that "the earth is the Lord's" is not expressed by landlords. If anything, the African intuition is that we belong to the land, not the land to us.[50] God's household is clearly not the only available image and is best balanced by all the others.[51] Suffice it to say that seeing the world through God's eyes as God's own beloved creation is very, very different from other ways of seeing it.

■ Bibliography

Barth, Karl. *Church Dogmatics 3.1: The Doctrine of Creation.* Edinburgh: T&T Clark, 1958.

Bavinck, Herman. *The Philosophy of Revelation.* London: Longmans, Green & Co, 1909.

Bosch, David J. "The Roots and Fruits of Afrikaner Civil Religion." In *New Faces of Africa: Essays in Honour of Ben (Barend Jacobus) Marais*, edited by Jan W. Hofmeyr and Willem S. Nicol, 14–35. Pretoria: Unisa, 1984.

Cloete, Daan, and Dirk J. Smit, eds. *A Moment of Truth: The Confession of the Dutch Reformed Mission Church, 1982.* Grand Rapids: Eerdmans, 1984.

Coetzee, Murray. *Die "Kritiese Stem" teen Apartheidsteologie in die Ned Geref Kerk (1900-1974): 'n Analise van die Bydraes van Ben Marais en Beyers Naudé.* Wellington: Bybelmedia, 2010.

Coetzee, Murray H., and Ernst M. Conradie. "Apartheid as Quasi-Soteriology: The Remaining Lure and Threat." *Journal of Theology for Southern Africa* 138 (2010), 112–23.

Cone, James H. "Whose Earth Is It Anyway?" In *Earth Habitat: Eco-Injustices and the Church's Response*, edited by Dieter T. Hessel and Larry Rasmussen, 23–32. Minneapolis: Fortress, 2001.

Conradie, Ernst M. *Christianity and a Critique of Consumerism: A Survey of Six Points of Entry.* Wellington: Bible Media, 2009.

———. "Could Eating Other Creatures Be a Way of Discovering Their Intrinsic Value?" *Journal of Theology for Southern Africa* 164 (2019), 26–39.

---

49. See Cone, "Whose Earth Is It Anyway?"; Zachariah, "Whose Oikos Is It Anyway?"

50. See the Machakos Statement, "The Earth Belongs to God," produced by an African Regional Consultation on Environment and Sustainability held at Machakos, Kenya, May 6–10, 2002, in preparation for the World Summit on Sustainable Development (WSSD) held in Johannesburg later that year. In a playful manner, where every statement is undermined by the next one only to complete a full circle, it stated that "In the household of God (oikos) the management of the house (economy) has to be based on the logic of the house (ecology). 1. In Africa today, it does not appear as if the earth belongs to God. Instead, it belongs to […] 2. God has entrusted the land and all its natural resources to all people to care for, keep and use it within communities. This requires a vision of sustainable communities […] 3. The land given to us by God does not only belong to the present community. It also belongs to our ancestors on whose contributions we build and whose memories we keep. It also belongs to the coming generations for whom we hold the land in trust and whose needs we should not compromise. 4. The land does not belong to us as people. Instead, we belong to the land. 5. The land does not belong to itself. Ultimately, it belongs to its Creator, the One who sustains the Earth, and who will finally restore it." For a discussion, see also Conradie, "A (South) African Land Ethic?"

51. See the extensive discussion on "oikotheology" in my *The Earth in God's Economy*, 221–46.

———. "Doing Creation Theology in the South African Context." *Journal of Systematic Theology* 1:6 (2022), 1–37.

———. *The Earth in God's Economy: Creation, Salvation and Consummation in Ecological Perspective*. Berlin: LIT, 2015.

———. *Saving the Earth? The Legacy of Reformed Views on "Re-Creation."* Berlin: LIT, 2013.

———. "A (South) African Land Ethic? The Viability of an Ecocentric Approach to Environmental Ethics and Philosophy." In *African Environmental Ethics: A Critical Reader*, edited by Munamato Chemhuru, 127–39. Heidelberg: Springer, 2019.

———. "Unravelling Some of the Theological Problems Underlying Discourse on 'Nature'." *HTS Theological Studies* 76:1 (2019), a6068. https://doi.org/10.4102/hts.v76i1.6068

———. "Views on Worldviews: An Overview of the Use of the Term Worldview in Selected Theological Discourses." *Scriptura* 113 (2014), 1–12. https://doi.org/10.7833/113-0-918

———. "What Is the Place of the Earth in God's Economy? Doing Justice to Creation, Salvation, and Consummation." In *Christian Faith and the Earth: Current Paths and Emerging Horizons in Ecotheology*, edited by Ernst M. Conradie et al., 65–96. London: T&T Clark, 2014.

———. "What on Earth Did God Create? Overtures to an Ecumenical Theology of Creation." *The Ecumenical Review* 66:4 (2014), 433–53. https://doi.org/10.1111/erev.12120

Conradie, Ernst M., ed. "The Ambiguity of Johan Heyns: Sitting on Bavinck's Left or Right Hand?" *Ned Geref Teologiese Tydskrif* 54:3&4 (2013), 6–18. https://doi.org/10.5952/54-3-4-396

———. *Creation and Salvation: Dialogue on Abraham Kuyper's Legacy for Contemporary Ecotheology*. Leiden: Brill, 2011.

———. *Creation and Salvation, Volume 1: A Mosaic of Essays on Selected Classic Christian Theologians*. Berlin: LIT, 2011.

———. *Creation and Salvation, Volume 2: A Companion on Recent Theological Movements*. Berlin: LIT, 2012.

———. "Ecumenical and Ecological Perspectives on the 'God of Life'." *The Ecumenical Review* 65:1, 1–2 (2013), 3–165. https://doi.org/10.1111/erev.12022

———. "How Are They Telling the Story?" *Scriptura* 97 (2008), 1–136; *Scriptura* 98 (2008), 137–243.

———. *Kontra-intuïtief: Om hierdie Wêreld as God se Skepping te Sien – 'n Gevaarlike én Verbysterende Belydenis*. Wellington: Bybelmedia, 2023.

Conradie, Ernst M., and Cynthia Moe-Lobeda, eds. *How Would We Know What God Is Up To? An Earthed Faith: Telling the Story Amid the "Anthropocene,"* Vol. 2. Cape Town: AOSIS, 2022. https://doi.org/10.4102/aosis.2022.BK295.

Conradie, Ernst M., and Lai Pan-chiu, eds. *Taking a Deep Breath for the Story to Begin … An Earthed Faith Volume 1*. Cape Town: AOSIS / Eugene: Wipf & Stock, 2021. https://doi.org/10.4102/aosis.2021.BK264

Faith and Order Commission. *Confessing the One Faith Study Document 153*. Geneva: WCC, 1991.

Fensham, Charles. "The Sacrament of the First Child of God: A Renewed Christian Eco-Imaginary." *Scriptura* 111 (2012), 323–32. https://doi.org/10.7833/111-0-16

Gunton, Colin. *The Triune Creator: A Historical and Systematic Study*. Grand Rapids Eerdmans, 1998.

Hall, Douglas John. *The Cross in Our Context: Jesus and the Suffering World*. Minneapolis Fortress, 2003.

Hefner, Philip. *The Human Factor: Evolution, Culture, and Religion*. Minneapolis: Fortress, 1993.

Jennings, Willie James. "Reframing the World: Toward an Actual Christian Doctrine of Creation." *International Journal of Systematic Theology* 21:4 (2019), 388–407. https://doi.org/10.1111/ijst.12385

Keller, Catherine. *Face of the Deep: A Theology of Becoming*. New York: Routledge, 2003.

Kinghorn, Johann, ed. *Die NG Kerk en Apartheid*. Braamfontein: Macmillan, 1986.

Lalu, Premesh, and Noëleen Murray, eds. *Becoming UWC: Reflections, Pathways and the Unmaking of Apartheid's Legacy*. Bellville: Centre for Humanities Research, 2012.

Lathrop, Gordon W. *Holy Ground: A Liturgical Cosmology*. Minneapolis: Fortress, 2003.

Liedke, Gerard, and Ulrich Duchrow. *Shalom: Biblical Perspectives on Creation, Justice and Peace*. Geneva: WCC, 1987.

Loubser, Bobby. *The Apartheid Bible: A Critical Review of Racial Theology in South Africa*. New York: Maskew Miller Longman, 1987.

Moltmann, Jürgen. *God in Creation: An Ecological Doctrine of Creation*. London: SCM, 1985.

———. *Sun of Righteousness, Arise! God's Future for Humanity and the Earth*. Minneapolis: Fortress, 2010.

Sittler, Joseph. *Evocations of Grace: The Writings of Joseph Sittler on Ecology, Theology and Ethics*. Edited by Peter Bakken and Steven Bouma-Prediger. Grand Rapids: Eerdmans, 2000.

Snyder, Howard. *Salvation Means Creation Healed: The Ecology of Sin and Grace*. Eugene: Cascade, 2011.

The Machakos Statement. "The Earth Belongs to God." *Bulletin for Contextual Theology in Africa* 8:2&3 (2002), 112-13.

Vaai, Upolu Lumā. "We Are Earth: reDirtifying Creation Theology." In *reSTORYing the Pasifika Household*, edited by Upolu Lumā Vaai and Aisake Casimira, 63-84. Suva: Pacific Theological College Press, 2023.

Van Keulen, Dirk J. "'We Zijn een Grap van God': Van Ruler over de Vreugde." In *Men moet Telkens Opnieuw de Reuzenzwaai aan de Rekstok Maken: Verder met Van Ruler*, edited by Dirk J. van Keulen, George Harinck, and Gijsbert van den Brink, 177-91. Zoetermeer: Boekencentrum, 2009.

Van Ruler, Arnold A. *This Earthly Life Matters: The Promise of Arnold Van Ruler for Ecotheology*. Edited by Ernst M. Conradie and translated by Douglas G. Lawrie. Eugene: Pickwick, 2023.

———. *Verzameld Werk, Deel III: God, Schepping, Mens, Zonde*, edited by Dirk Van Keulen. Zoetermeer: Boekencentrum, 2009.

Veenhof, Jan. *Nature and Grace in Herman Bavinck*. Translated by Albert M. Wolters. Sioux Center: Dordt College Press, 2006.

Wallace, Mark I. *When God was a Bird: Christianity, Animism, and the Re-Enchantment of the World*. New York: Fordham University Press, 2019.

Welker, Michael. *Creation and Reality*. Minneapolis: Fortress, 1999.

Westhelle, Vítor. "Creation Motifs in the Search for a Vital Space: A Latin American Perspective." In *Lift Every Voice: Constructing Christian Theologies from the Underside*, edited by Susan Brooks Thistlethwaite and Mary Engel Potter, 146-58. Maryknoll: Orbis, 1998.

———. *Eschatology and Space: The Lost Dimension in Theology Past and Present*. New York: Palgrave McMillan.

Zachariah, George. "Whose Oikos Is It Anyway? Towards a Poromboke Eco-Theology of 'Commoning'." In *Decolonizing Ecotheology: Indigenous and Subaltern Challenges*, edited by S. Lily Mendoza and George Zachariah, 205-22. Eugene: Wipf & Stock, 2021.

The Story of a Place in the North: Natural Disasters within God's Good Creation—A Lutheran Perspective

Arnfríður Guðmundsdóttir[1]

My home country, Iceland, is sometimes called *the land of ice and fire*. The name is particularly fitting in times of a warming climate, as we watch the glaciers melting, and feel the earth shaking beneath our feet. Melting glaciers are causing the land to rise, which is likely to result in more frequent earthquakes and volcanic eruptions. Since 2010, there have been numerous eruptions, some gaining worldwide attention because of the effect they had on international air traffic. In 2021, an eruption took place midway between our capital, Reykjavík, and our international airport in Keflavík.

---

1. Arnfríður Guðmundsdóttir is a professor of systematic theology, Faculty of Theology and Religious Studies, University of Iceland. She is registered as a coresearcher at the University of the Western Cape for the project on "An Earthed Faith: Telling the Story amid the 'Anthropocene'."

How to cite: Conradie, EM 2024, "The Story of a Place in the North: Natural Disasters within God's Good Creation—A Lutheran Perspective", in EM Conradie & WJ Jennings (eds.), *The Place of Story and the Story of Place*, in An Earthed Faith: Telling the Story amid the "Anthropocene", vol. 3, AOSIS Books, Cape Town, pp. 101-119. https://doi.org/10.4102/aosis.2024.BK355.06

The Story of a Place in the North: Natural Disasters within God's Good Creation

The volcano had been sleeping for 800 years. It erupted again in 2022, but then the eruption lasted only a few weeks, while it had lasted six months the previous year. Fortunately, this particular volcano is located in an uninhabited area, but the closeness to a densely populated district and our main airport caused a major concern. Still, not all eruptions happen in the middle of nowhere. Back in 1973, an eruption happened in the middle of a town in Heimaey, which belongs to the Westman Islands, just south of Iceland. The entire island was evacuated overnight, but 5,000 people lived there at the time. Before the eruption was over, around 400 houses had been buried under the lava. In addition to the imminent danger of volcanic eruptions, avalanches and mudslides, together with severe weather, pose a constant threat to a small nation living on a fairly small island just south of the Arctic Circle.

While stories of natural disasters form an integral part of who we are as Icelanders, they do not necessarily prevent us from seeing what is good about our island. I will name just a few examples. Firstly, there are the natural resources, something all of us benefit from on a daily basis, resources such as pure water, geothermal water, and prolific fishing grounds. Secondly, there is the intact nature. Because a very limited part of the island is habitable, the biggest part of the inland is uninhabited territory, much appreciated for its beauty and serenity. If I take into account only the beautiful landscape, and the great assets it has to offer, it is not difficult to see my home *as the creative work of God's love*. But that can in fact become challenging when nature turns against us, and our lives are threatened by various natural disasters. When faced with life-threatening events, like volcanic eruptions or avalanches, or a heavy snowstorm in a rural area, where no shelter is in sight, do we still see our home as God's creation, a place God has provided for us? Or does it then look more like a godforsaken country, where we are left on our own to fight against evil forces, with no help in sight?

My focus in this essay will be on different theological responses to critical situations, past and present. I have chosen an event from the eighteenth century as an example of a natural disaster which has made a lasting impression in our society. We learn about it in school as kids, and it is carefully commemorated in our literature. This is one of the most destructive events in our history, killing more than 20% of the population, and a significant part of the livestock at the time. The reason why it is so deeply ingrained in people's memory is because of the detailed description which has been preserved of what happened. The person who provided the account was a pastor, who was not only interested in giving accurate information about the developments but was even more preoccupied with the theological interpretation of the event. His account provides important background for my inquiry into the language of lament and what that has to offer to people who are caught in life-threatening situations, looking for help where help is nowhere to be found. But before I turn to more or less helpful responses to traumatic experiences, I will start by describing the place where I was born and raised.

■ My Place of Origin

Settlement in Iceland goes back to the second half of the ninth century CE, when people started to come over from Norway, looking for a place to live, away from the jurisdiction of the Norwegian king. Irish monks were most likely living in Iceland at that time. Very limited information is available about the monks, how many they were, and their whereabouts.[2] Old Norse religion was the dominant religion until the year 1000, when it was decided in the Icelandic parliament that Christianity should become the official religion. The presence of Christian slaves, brought to Iceland by the settlers, is likely to have helped make the transition a peaceful one, as their responsibility was often to care for and educate children. At the time of the Reformation, Iceland belonged to the Danish Crown, and it was the decision of the Danish king that the Icelandic church should no longer be a part of the Roman Catholic Church. From the middle of the sixteenth century, the church belongs to the Lutheran tradition. The Evangelical Lutheran Church of Iceland is a national church, as has been declared in the Icelandic Constitution since 1874. Even if the majority of the population still belongs to the national church, a growing number of citizens belong to other dominations or different religions.[3]

Iceland is an island in the Atlantic Ocean, located between North America and Europe.[4] It belongs to the Arctic region, and the Arctic Circle goes through an island which belongs to Iceland and is located just north of it. It is the least populated country in Europe, with a little over 370,000 inhabitants. Only 20% of the island (the coastline) is habitable.[5] The island is one of the youngest landmasses on the planet and home to some of the

---

2. See "Irish Hermit Monks Were Said To Have Inhabited Iceland Before Scandinavian Settlers," *The Historian's Hut*, July 20, 2019, https://thehistorianshut.com/2019/07/20/irish-hermit-monks-were-said-to-have-inhabited-iceland-before-scandinavian-settlers/.

3. Guðmundsdóttir, "The Lutheran Church in Iceland," Christus Victor Lutheran Church, last accessed September 15, 2022, https://cvlutheran.org/reformation-iceland.

4. Guðmundsdóttir, "The Fire Alarm Is Off," 141–44.

5. Iceland is an island of 103,000 km² (40,000 square miles), the second largest island in Europe, following Great Britain, and the 18th largest island in the world. The coastline is 4970 km, and Iceland maintains a 200 nautical-miles exclusive economic zone. See last accessed June 17, 2023, https://www.visiticeland.com/article/iceland-in-numbers.

world's major volcanoes.[6] Because it is still an active volcanic area, its landscape is literally still being formed. It is divided by the Mid-Atlantic Rifts, where the two plates, the Eurasian and North American tectonic plates, are slowly moving apart. At the breach, it is growing about two inches (5.8 cm) every year.

More than 10% of Iceland is covered by glaciers, which are melting as a result of rising temperatures.[7] Vatnajökull, in the southeastern part of the country, is the largest glacier in Iceland, but is also the largest glacier in Europe. Shrinking glaciers are causing the "land" to rebound from the Earth's crust—a process that is happening at a pace much faster than scientists had predicted. Glaciers are melting so swiftly that parts of Iceland are rising as much as 1.4 inches (35 millimeters) a year. One example is Sólheimar Glacier, which retreated just over 3,000 feet (914.4 meters) from 2000 to 2015. Where ice existed before at the glacier tongue, there is now a large glacier lagoon.[8] As the weight of the glaciers has lessened, the rising land is causing increased volcanic activity and earthquakes. There are between 30 and 40 active volcanoes, which means that volcanic eruptions and earthquakes are fairly common. Volcano activity is a steady reminder of how dependent we really are on nature. There was a major volcanic eruption in 1875, which affected the lives of people all over Iceland. At that time, the population of Iceland was approximately 70,000. During the second half of the 1800s, somewhere between 16,000 and 20,000 Icelanders emigrated, mainly to North America.[9] Hence, by the end of the nineteenth century, up to 25% of the population had emigrated because of harsh weather and natural disasters.

I was born and raised in a small fishing town in the northern part of Iceland. It is actually the northernmost town on the island, located in a narrow fjord, surrounded by steep hills. When I was a kid, a very steep

---

6. See last accessed June 17, 2023, https://www.britannica.com/topic/worlds-major-volcanoes-2226816.

7. An Icelandic glaciologist, Guðfinna Th. Aðalgeirsdóttir, one of the contributors to the Intergovernmental Panel on Climate Change (IPCC) report from 2021, believes that if things continue as they are, "all but the biggest one" will be gone by the end of this century. "The Entrance to the 'Centre of the Earth' May Disappear within Decades." *United Nations, Regional Information Centre for Western Europe*, https://unric.org/en/the-glaciers-in-iceland-may-disappear-within-decades/; "Guðfinna Aðalgeirsdóttir, Professor at Faculty of Earth Sciences Is One of the Lead Authors of the Report of the Intergovernmental Panel on Climate Change," *The University of Iceland, The Science Institute*, last accessed September 18, 2022, https://raunvisindastofnun.hi.is/frettir/2021_08_10/gudfinna_adalgeirsdottir_professor_at_faculty_of_earth_sciences_is_one_of_the.

8. Richard Engel, Charlotte Gardiner, and Kennett Werner, "How to Save a Glacier: Iceland's Scientists Offer Hope with Carbon Capture Technology," *NBC News*, September 19, 2019, https://www.nbcnews.com/news/world/how-save-glacier-iceland-s-scientists-offer-hope-carbon-capture-n1052281.

9. See "New Land, New Life". *The Icelandic Emigration Center*, last assessed September 18, 2022, https://www.hofsos.is/general-8.

mountain road was the only way to get to and from the town on land. The road was only open during the summer. In winter, the sea was the only way to travel to and from our town. It goes without saying that people felt framed in. For most people, it was a comfortable feeling, a source of security, to be embraced by the mountains, while others felt trapped, especially when virtually everything was covered in snow. But for all of us, the danger of avalanches was real, something which was always lurking at the back of our minds. It was part of our growing up to listen to stories of avalanches destroying houses and infrastructures, sometimes fatally, as in April 1919 when eighteen people were killed.[10] When things started to look precarious, people who lived in the most dangerous areas had to leave their houses until things were deemed safe again. Even if I did not live in one of those houses, it affected all of us when it happened. It made it more real, how vulnerable we were vis-à-vis the strong forces of nature. One of the most compelling memories from my childhood is when a boy from my class in first grade came to school after an avalanche had hit his house. An oil tank had been located by the house, facing the mountain, and the avalanche had taken the tank with it through the house, leaving it in the garden on the other side. Fortunately, the whole family survived, as their bedrooms were in a different part of the house.

In addition to avalanches and landslides, experiences of volcanic eruptions and earthquakes are an important part of our story. Ever since Christianity became the official religion of the Icelandic nation, around the year 1000, it has been an ongoing challenge for the people to make sense, in light of their faith, of frequent natural disasters. There are ample examples of people who have told their stories and witnessed about their faith in God, their Creator, who has given them courage and strength in challenging circumstances. It has been common for people to see the country and its resources as gifts from God the Creator, and a sign of God's grace, and natural catastrophes as a manifestation of God's anger and revenge. It is safe to say that respect for nature is deeply rooted in the nation's psyche. But for those who have survived natural disasters, it has been an enormous challenge to work through their experiences in order to be able to go on with their lives.

■ A Pastor Serving God and Distressed People in Catastrophic Circumstances

Volcanic eruptions have made a lasting impact on life on our island as far back as written sources go. The biggest eruption in recorded history

---

10. Vala Hafstað, "Siglufjörður Avalanche Remembered," mbl.is, April 13, 2019, https://icelandmonitor.mbl.is/news/nature_and_travel/2019/04/13/siglufjordur_avalanche_remembered/.

happened toward the end of the eighteenth century. It occurred in the southeastern part of Iceland, when a fissure seventeen miles (27 km) long opened up in Laki (also known as Skaftá Fires in Icelandic).[11] It was one of the largest volcanic eruptions recorded worldwide, and it made a notable impact in the Northern Hemisphere.[12] It lasted from June 1783 to February 1784, with tragic consequences for all living creatures – not only in the vicinity but all over Iceland. As a result of gases from the eruption, fields, meadows, and ponds were poisoned, killing a large portion of animals in the country, especially cattle, sheep, and horses. The Skaftá Fires were followed by severe famine, and by 1785, roughly 20% of the Icelandic population had died from hunger, malnutrition, or diseases.[13]

Detailed information about the Laki eruption has been reserved in the writings of Rev. Jón Steingrímsson (1728–1991), a local pastor and provost in the southeast region. Steingrímsson's description appears in his autobiography, personal letters, official records, and in his treatise about the fires, in an English translation entitled *Fires of the Earth: The Laki Eruption 1783–1784*.[14] If not for his writings, only a scattered picture of this major event in our history would be known. Steingrímsson had been captivated by volcanic eruptions from an early age and had already written an account of an earlier eruption of a volcano named Katla in 1755.[15] This is worth noting because it explains his familiarity with the subject matter when it came to his documentation of the Laki Eruption.[16] Steingrímsson was not only interested in volcanic activities but in science in general, especially medical science. Later in life, he became an amateur physician and treated people under the supervision of the surgeon general, who was a close friend of his. Not only did he prescribe medications, but he also operated on people on several occasions. At the time, physicians were few and far apart, which made Steingrímsson's medical work much appreciated. While he was not able to go abroad for university education, he was nevertheless well educated. He attended school in Hólar, one of the two

---

11. Its impact reached most parts of Europe, Canada, Alaska, and even as far as the borders of China. See, for example, Katrin Kleeman, "The Laki Fissure eruption, 1783–1784," *Encyclopedia of the Environment*, January 14, 2020, https://www.encyclopedie-environnement.org/en/society/laki-fissure-eruption-1783-1784/.

12. See Erik Klemetti, "Local and Global Impacts of the 1783–84 Laki Eruption in Iceland," June 7, 2013, https://www.wired.com/2013/06/local-and-global-impacts-1793-laki-eruption-iceland/.

13. See Kleeman, "The Laki Fissure Eruption."

14. The treatise was first published in Icelandic in 1907. See "*Skráningarfærsla handrits*," last accessed September 25, 2022, https://handrit.is/manuscript/view/is/Lbs04-1552/0#mode/2up.

15. Katla is one of the most active and most dangerous volcanoes of Iceland, also in the southeastern part of the country. It erupts on average every 50–100 years. See "Katla, Volcano, Iceland," *Brittanica*, last accessed September 25, 2022, https://www.britannica.com/place/Katla.

16. Rafnsson, "*Um eldritin*," 246.

bishoprics in Iceland at the time. When he graduated, he was qualified to become a pastor within the Evangelical Lutheran Church.

Even if Steingrímsson was not an ordinary pastor by any means, he nevertheless dedicated his life to his pastoral duties. He was a pastor who stood by his people, regardless. This became obvious during the trials and tribulations following the eruption. While every other pastor in the vicinity had left his congregation and fled to a less dangerous part of the country, Steingrímsson stayed behind. He was convinced that he had been called to serve God's people, and he wanted to stay true to his vocation. It may look like a contradiction, but it is nevertheless true that while he was sure the eruption was in agreement with God's will, he did everything he could to help his people. Providing pastoral care in the midst of a volcanic eruption, Steingrímsson did not limit his care to counseling, even if he provided it generously to desperate souls. He helped people in every way he could. He fed the hungry and took in people who had no place to stay. Despite the immediate danger, worship services were held in church buildings, as well as funerals for the people who died of afflictions related to the eruption. From 1783 to 1785, more than one-third of his parishioners passed away. Sometimes, there were no healthy people available who were strong enough to dig graves in frozen ground for all the dead bodies. This is why some of them had to be buried in mass graves, up to ten people in a single grave. But it was important to Steingrímsson that nobody was buried without a coffin.[17] In the end, he himself became both physically and mentally sick from exhaustion. In his autobiography, he admits that there were times he suffered from suicidal thoughts.[18] According to a 21st-century diagnosis, he was most likely suffering from post-traumatic stress disorder and would have needed the type of support that was not available in his time.[19]

■ "The Destructive Fires of God"[20]

The most famous writing of Rev. Jón Steingrímsson is his autobiography, the first autobiography written in Icelandic. Steingrímsson's story is a one-of-a-kind portrayal of the role of a pastor during an emergency situation. It is a detailed testimony of an eighteenth-century person coming of age and how he managed to work his way out of sheer poverty and through great hardship. At the same time, it is an important account of a struggling nation,

---

17. Steingrímsson, *Fires of the Earth*, 79.
18. Steingrímsson, *A Very Present Help in Trouble*, 212–13.
19. Hugason, "*Sjálfsvitund á barmi taugaáfalls*," 233.
20. Steingrímsson, *Fires of the Earth*, 12.

which was at the time a part of the Danish Crown, and therefore totally dependent on a foreign power. The Laki eruption and the famine toward the end of the 18th century were nothing less than natural catastrophes and a real threat for future habitation in Iceland. Help was limited and late to arrive from Denmark, and Icelandic authorities did not have what they needed to provide help for their people.

In a preface to his book, Steingrímsson states that his intention was to tell his five daughters and their descendants who he was and where he came from.[21] His meticulous style and utmost sincerity show his intention to tell the whole story and leave nothing behind. Whether he really did so is unknown. To some extent, the book resembles Augustine's *Confessions*, or any other confessional text, with the emphasis on God's mercy and providence despite one's sinful nature and shortcomings.[22] More than anything, he was witnessing about his faith in God, who created him and chose him for this particular task.

Steingrímsson kept notes during the eruptions and later collected data about the dire ramifications. He started writing the treatise five years after the eruptions started. His treatise is both a detailed description of a natural disaster and a theological interpretation of the event. Hence, it is a historical and scientific account, but at the same time, it is a religious testimony. As it was written at a time in history when the Enlightenment was starting to make an impact within different disciplines, it was still not seen as necessary to make a clear distinction between natural science and theology. For Steingrímsson, reason and faith were not antitheses, as they were both God's good gifts. While he was influenced by the strong emphasis of his time on reason, Steingrímsson remained true to the orthodox Lutheranism he knew from his childhood and studied as a young man.

Steingrímsson gives several reasons why he decided to write his account. Firstly, he considered it his duty to testify to God's grace and give thanks for God's protection through troubling times. Secondly, he meant to educate his people and help them understand the will and wonders of God. He wanted to respond to questions like: Why did it happen? Why did God allow it to happen? Steingrímsson was convinced that it was all for a reason, namely God's reason, which was to punish his people for showing apathy and being unfaithful during the prosperous period leading up to the eruption.[23] When God was not able to get through to them by using more

---

21. Steingrímsson, *A Very Present Help in Trouble*, 22-23.

22. Hugason, "*Honum sje eilíft of og æra fyrir strítt og blítt*," 43.

23. Steingrímsson, *Fires of the Earth*, 14-16. It is very interesting to compare Steingrímsson's account to the religious interpretations of the people who experienced Cyclone Idai in Zimbabwe in 2019. See Chirongoma and Chitando, "What Did We Do to Our Mountain?" 69-82.

temperate means, God decided to use fire. Even if fire was by nature a good thing, the pastor was convinced that it could become a tool for the Creator to chastise God's people.[24] But because God was "the source of good and bad," God did not leave the people on their own and made sure to give everybody a chance to flee from the fire.[25]

What happened in this particular eruption was, according to Steingrímsson, in no way a unique event; rather, it was similar to what had happened so many times before and had been attested to in annals and other writings in the past.[26] As before, God did not mean to eliminate all populated areas and would, in the end, turn condemnation into blessings. When people finally saw the rainbow again, right before a worship service on a Sunday morning, months after the beginning of the eruption, they took it as a sign from God and were convinced that God would eventually protect them from the fire.[27] In retrospect, Steingrímsson concluded that the hardship had been for the best, as people had learned not only to appreciate what God had given them but also to pursue humility and patience.[28]

The most celebrated incident from Steingrímsson's treatise about the fire was the so-called fire service. It was a church service he celebrated on July 20, 1783, when the fires had been burning for about six weeks. At that time, two out of three churches in his parish had already burned down. People who came to church that morning thought this might be their last visit, given that this church would also disappear soon. According to the pastor himself, nobody was afraid once they were inside the church, nor were they impatiently waiting for the service to be over. Instead, they kept calm during the service and prayed for God's mercy. After it was over, they all stepped out to see what had happened while they were inside the church. Instead of continuing in the direction of the church, the lava had stopped and started to pile up at a riverbank just outside of the church.[29] Steingrímsson describes how the water managed to cool down the lava and stop the flowing. At the same time, he was sure that nothing less than

---

24. Steingrímsson, *Fires of the Earth*, 11.

25. Steingrímsson, *Fires of the Earth*, 29.

26. Steingrímsson, *Fires of the Earth*, 13.

27. Steingrímsson, *Fires of the Earth*, 60.

28. Steingrímsson, *Fires of the Earth*, 89–90.

29. This "technique" was used in order to steer the flow of the lava during the eruption in Heimaey in 1973, when people experimented with cooling the lava down by spraying seawater on the edges. See "Lava-Cooling Operations During the 1973 Eruption of Eldfell Volcano, Heimaey, Vestmannaeyjar, Iceland, U.S. Geological Survey Open-File Report 97-724," U.S. Geological Survey, last accessed October 1, 2022, https://pubs.usgs.gov/of/1997/of97-724/32html/lavaoperations.html.

a miracle had happened. Rather than choosing either/or (either nature or God, reason or faith), he seems to have no problem with seeing both at work at the same time. Seeing how their church had been saved from destruction gave people hope and reason to give thanks to God, who had saved them from great danger.[30]

To be Created and Cared for by a Loving God

To confess faith in God, "the creator of heaven and earth, all that is seen and unseen," is the same as to declare that there is nothing outside the scope of God's creativity. The idea of a creation implies, firstly, that everything that exists originates from the Creator and, secondly, that all parts of creation are related to each other. What we learn from the creation stories in the first and second chapters of the book of Genesis is that God is intimately connected with all creatures, communicates with them, and bestows God's blessings upon them.

In his explanation of the first article of the Apostolic Creed in his *Large Catechism* (1529), Martin Luther explains what it means to talk about God, *our Creator*, in the following way:

> What is meant by these words or what do you mean when you say, "I believe in God, the Father almighty, creator," etc.?
>
> Answer: I hold and believe that I am God's creature, that is, that God has given me and constantly sustains my body, soul, and life, my members great and small, all my senses, my reason and understanding, and the like; my food and drink, clothing, nourishment, spouse and children, servants, house and farm, etc. [...]
>
> Moreover, we also confess that God the Father has given us not only all that we have and what we see before our eyes, but also that God daily guards and defends us against every evil and misfortune, warding off all sorts of danger and disaster. All this God does out of pure love and goodness, without our merit, as a kind father who cares for us so that no evil may befall us.[31]

What I take from these words is a picture of God who is not distant and indifferent about God's creation but present, God with us, deeply immersed in our created reality. It is an important contrast to an overemphasis on God as transcendent, sovereign, distant, and separated from the world, which so often has dominated the image of God within the Christian tradition and

---

30. Steingrímsson, *Fires of the Earth*, 48–50. Chirongoma and Chitando also tell about church buildings which survived the calamity caused by Cyclone Idai. They write: "The general conception among the people who survived the cyclone in Chimanimani is that the Lord God mysteriously saved the church buildings to demonstrate God's glory, as well as to provide sanctuary for the survivors." (Chirongoma and Chitando, "What Did We Do to Our Mountain?" 72).

31. Luther, *The Large Catechism*, 354.

so easily overshadows God's love and God's presence within creation. The term used to describe this understanding is *panentheism* [literally: "all in God"], a term which is meant to emphasize the fact that "God can be neither conflated with the created world nor separated from it."[32] The relationship between the mother and the fetus in utero is a powerful symbol of the intimate relationship between creator and creation.[33]

■ Prayers of Lament

When something terrible happens, like the Laka eruption toward the end of the eighteenth century, people of faith are likely to ask questions like: Why did it happen? And: Why did God allow it to happen? These are among the questions Steingrímsson intended to address in his writings. But are his answers the only options? To connect sin to suffering and evil, as he does, is in fact very common within the Christian tradition. But it is not the only alternative. The best-known example is from the Book of Job, which can be seen as a radical protest against the hardship he is experiencing. When Job regrets being born (Job 3:11–13), his friends try to convince him to regard his miseries as an indication of God's punishment. They even want him to be grateful for what he is going through (Job 5:17–18). But Job does not want to have any of that and chastises his friends for being "miserable comforters" (Job 16:2). Instead, he claims the right to speak out and protest:

> Therefore I will not keep silent; I will speak out in the anguish of my spirit, I will complain in the bitterness of my soul. (Job 7:11, NRSVue)

Job's response is certainly not unique within the Bible, but it belongs to prayers of lament, which are different from penitential prayers or prayers of confession, where the focus is on guilt and remorse. Katheleen M. O'Connor writes in the introduction to her commentary on Job about the characteristics of lamentations:

> These are prayers in the form of complaint. The one who prays cries out to God in pain, describes various forms of affliction, and seeks redress. Laments are a form of truth-telling to God. They open up rather than deny suffering, and they present to God the affliction of the one who prays and demands action. In Old Testament laments, anger at God and fury close to blasphemy are instruments of fidelity because they keep the relationship with God alive in the midst of suffering. They are acts of faith that God cares for the afflicted and can bear to

---

32. Moe-Lobeda, Cynthia D., *Resisting Structural Evil*, 143.

33. The feminist theologians Elisabeth Johnson and Sallie McFague, pioneers of feminist ecotheology, have used the concept "panentheism" in their writings. See Johnson, *She Who Is*, 231; McFague, *The Body of God*, 149; and "Reimagining the Triune God for a Time of Global Climate Change," 110–11.

hear the truth. By using laments, Job grows strong and courageous as he clings to God under the worst conditions.[34]

To believe that we are created and cared for by a God who is, in line with Luther's description in his *Large Catechism*, like "a kind father who cares for us so that no evil may befall us," rings true to our experience when things are going well, and we do indeed feel loved and cared for. But when suffering and evil take over our lives, our understanding of who God is does not always coincide with our experiences. Then these questions will likely become ours: *Where* is my God? and *Who* is my God?[35]

■ When the Creative Work of God's Love Is Destroyed by Natural Disasters

Traumatic experiences caused by violence[36] or life-threatening illnesses,[37] for example, are likely to trigger questions regarding people's faith. Questions related to suffering and evil are often about the nature of God and how God can possibly be loving, just, and all-powerful and still remain untouched by, or at least passive vis-à-vis, that which is causing their suffering and pain. Traditionally, such questions are called *theodicy* questions. In his foreword to Deanna Thompson's compelling book *Glimpsing Resurrection. Cancer, Trauma, and Ministry*, Willie James Jennings denounces theodicy questions as "bad questions," because he thinks they "drive people into endless, exhausting searches for answers they will never find, because theodicy questions are always self-enclosed."[38] Jennings maintains that asking questions is important and that Christian theology, together with the Christian community, are meant "to help people ask the right kinds of questions."[39] What characterizes good questions, according to Jennings, is that they are:

> [/]ntense and personal, urgent and angry, and relentless, always wanting to hear and know, see and sense God responding. Such questions begin with the real God: a God who is touched with the infirmities of the creature and the creation, acquainted with grief, familiar with sorrow and with very bad news.[40]

---

34. O'Connor, *Job*, 6.

35. See Guðmundsdóttir, "Talking About a Gracious God," 238.

36. See for example Serene Jones's book on *Trauma and Grace: Theology in a Ruptured World* (2009).

37. See for example Deanna A. Thompson's book, *Glimpsing Resurrection: Cancer, Trauma and Ministry* (2018).

38. Jennings, "Foreword," ix.

39. Jennings, "Foreword," x.

40. Jennings, "Foreword," xi.

I find the distinction between bad theodicy questions and "good questions" particularly helpful in the context of natural disasters, when people have survived traumatic experiences. I think it makes all the difference for people who have lost their homes in a volcanic eruption, literally in their backyard, or an avalanche tearing their houses as well as their communities apart, that they get to ask those *intense and personal, urgent and angry* questions. What they have experienced is that their *place*, as they have known it all their life, is no longer there but gone for good.

■ Eruption in Heimaey

Fifty years ago, a volcano started erupting in Heimaey, a small island south of Iceland.[41] Researchers estimate that the last eruption there happened 5,000 years ago. When people went to sleep on the evening of January 22, 1973, nobody expected that anything extraordinary would take place in the middle of the night. All the ships were in harbor because of a severe storm the day before. The weather had calmed down when the eruption started. Some 5,300 people lived in the town at this time, and within five hours, most of them were gone, by boat or by plane. Domestic animals had to be euthanized, or else they would be killed by lava or gas pollution. The eruption lasted for five months, and at that time around 25% of the houses had been destroyed. One person was killed by toxic gas from the eruption, which had accumulated in a basement.[42]

While the eruption itself was over in five months, it took years to rebuild the town. Still, it took people much longer to recover. The mountain they had known all their life, and which was the first thing they saw when they looked out of their windows in the morning, had, without any warning, started spewing ashes, lava, and dangerous gas pollution over their town, their home. Even if people were relieved that it was possible to evacuate the island in only a few hours in the middle of the night, it was unavoidable to raise "what if" questions, like "What if the weather had still been bad, and the storm had prevented the airplanes from landing?" or "What if the ships had not been in harbor?" In retrospect, people frequently indicate that it was no less than a miracle that no one was injured or killed before or during the evacuation. "Miracle" is, in fact, a word regularly used to describe what took place at the beginning of the eruption. After all, what happened to the more than 5,000 people that night was that they were uprooted and scattered around Iceland. They had become refugees in their own country.

---

41. See "50 years since the eruption in Heimaey," mbl.is, January 23, 2023, https://icelandmonitor.mbl.is/news/news/2023/01/23/50_years_since_the_eruption_in_heimaey/.

42. Sveinsson, *Útkall*, 163.

Life for those who stayed behind in order to rescue whatever they could in terms of buildings and infrastructure was also taxing. Soon, they started trying to steer the lava stream by spraying seawater on the flowing lava. Eventually, they managed to preserve the harbor.[43] Reflecting on their experience, to some, it felt like they were waging war with nature.[44] A policeman who stayed behind recounts how he felt during the first days and weeks when people worked without much rest, practically around the clock. He writes:

> I was full of rage and helplessness. No safety. No normal life. "God, where are you? Don't you have time to stop by? Is the enemy going to destroy everything?"[45]

The pastors in the local church followed the people to the mainland. Two months into the eruption, the senior pastor decided to go back and offer a worship service, a "fire service," in line with what his colleague Jón Steingrímsson had done back in the late eighteenth century. In order to get to the church, loads of ashes and tephra had to be removed. Everyone who was able to go went to church. People remember a profound and touching experience. The pastor prayed to God that the lava flow would not break through a barricade that had formed in the crater. But in the wake of the service, the eruption became even more powerful than before, which was heartbreaking for those who had hoped God would indeed answer their prayer.[46] These were trying times, life-changing for those who experienced the huge damage to their place. One person who stayed in Heimaey during the eruption told an author writing about the event: "We will never be the same, the way we look at life, is changed forever."[47]

■ Avalanches in the Westfjords

In many ways, it is hard to compare what happened in Heimaey in 1973 with the devastating and deadly avalanches that fell on two villages in the Westfjords (in the northwestern part of Iceland) in January and October of 1995. The harmful events originated in the mountains just above the villages.[48] But while the volcanic eruption in Heimaey lasted for five months, the events in Súðavík and Flateyri happened in the twinkling of an eye.

---

43. "Lava-Cooling Operations During the 1973 Eruption of Eldfell Volcano, Heimaey, Vestmannaeyjar, Iceland U.S. Geological Survey Open-File Report 97-724," U.S. Geological Survey, last accessed October 1, 2022, https://pubs.usgs.gov/of/1997/of97-724/edintro.html.

44. Sveinsson, *Útkall*, 189.

45. Sigurðsson, *Undir gjallregni*, 31 (translation AG).

46. Sigurðsson, *Undir gjallregni*, 106.

47. Sveinsson, *Útkall*, 198 (translation AG).

48. Chirongoma and Chitando; "What Did We Do to Our Mountain?"

Still, the biggest difference is how fatal the incidents in the northwest turned out to be.

Súðavík is a small village, where a little over 200 people lived in January of 1995 when the avalanche fell and destroyed or seriously damaged over 20 houses, more than 30% of all the houses in the area. Forty-seven people lived in the houses that were hit by the avalanche. Nineteen escaped from beneath the snow on their own, fourteen were rescued, and fourteen died.[49] The rescue operations continued for 36 hours, but a relentless blizzard made the rescue operations extremely difficult. Close to 300 people lived in Flateyri at the time of the avalanche in October of 1995. Thirty-three houses, with 54 people, were hit when the avalanche fell in the middle of the night. Thirty-five were saved and 20 died.[50] As in Súðavík, rescue work was stopped after 36 hours. Some important lessons had been learned from Súðavík, which helped in administering the rescue operations in Flateyri. The experiences from both places have also made people more aware of the long-lasting impact that such trauma has on people. Many individuals have told their stories, which has given the rest of us some understanding of what it takes to work through the suffering and pain they experienced in the aftermath of these catastrophic events.

There are several things I find particularly noteworthy about the avalanches in Súðavík in January 1995 and Flateyri ten months later. First and foremost is the indescribable sorrow and pain people experienced. It was a sheer tragedy, especially for those who survived or were related to those who died, but so many more were affected by it. Iceland is a close-knit nation, especially when natural disasters occur. Just hours after the events, there were prayer meetings in churches all over the country.[51] Icelandic people are not known for being church-going people, but when something terrible happens, they do show up. Open churches not only gave people a chance to pray together, but they also provided a forum to meet, to express their pain and suffering, and to receive comfort and support. The solidarity that the entire nation showed in the aftermath of the avalanches was very important for those who were hardest hit.

---

49. Bernhardsdóttir, "Learning from Past Experiences," 7. In the wake of the avalanche in Súðavík, up to 1,000 people had to leave their homes in various places in Westfjords because of a great danger of avalanches (Fjeldsted, *Þrekvirki*, 171). A father and son were caught in an avalanche two days later, south of Súðavík. The son was rescued after twelve hours and survived against all odds. His father was dead when the rescue team found him (Fjeldsted, *Þrekvirki*, 199–200).

50. Bernhardsdóttir, "Learning from Past Experiences," 36.

51. Eiríksdóttir and Johnson, *Nóttin sem öllu breytti*, 187.

■ "Christs to One Another"

When faced with natural calamities of some sort, the question of God's power becomes a pressing one. Feminist theologians have reacted to the traditional understanding of God's power, coming up with different ways of interpreting it. Instead of understanding power in terms of dominion or coercion, Wendy Farley has suggested that we interpret it as *compassion*.[52] Elizabeth Johnson builds on Farley's idea of *divine compassion* when she describes God's active solidarity with those who suffer as God's "compassion poured out."[53] God's compassion is consequently mediated through humans, who show compassion to those in need. This is what Martin Luther calls becoming "Christs to one another."[54]

From testimonies of people who have survived natural disasters, it is obvious how important it has been to them to receive support and understanding from other people, be they family, friends, rescue-workers, professionals (including pastors), or perfect strangers. Just to know that somebody cares has made all the difference to them. It has also been crucial to be able to talk about their experiences, including the experience of God's silence or absence, as they have been confronted with the consequences of what happened. This is often very challenging to hear, which might prevent people from being honest about their experiences, even in church. "Prayers of lament" do not play a significant role within the Christian liturgy, with serious consequences, according to Kathleen Billmann and Daniel Migliore. They write:

> [/]nstead of providing space for protest and grief, what churches often offer are worship services that are "unrelentingly positive in tone." The exclusion of the lament screens out people who find the services shallow or harmful and provides not theological and liturgical way to come to terms with disturbing experiences.[55]

The fact that people do come to church when something as terrible as a volcanic eruption starts in the middle of their town or when avalanches destroy big parts of their villages discloses their need for a place not only to feel support, but also to express what they have experienced. In those situations, it is crucial that the church become a place where people experience compassion, a place where they are free to speak their mind and call the thing "what it actually is."[56]

---

52. Farley, *Tragic Vision and Divine Compassion*, 86.

53. Johnson, *She Who Is*, 246. See also Guðmundsdóttir, *Meeting God on the Cross*, 146–48.

54. Luther, "The Freedom of a Christian," 368.

55. Billman and Migliore, *Rachel's Cry*, 14. See also Brueggemann and Linafelt, 317–318.

56. Luther, *Heidelberg Disputation*, 53. See also Guðmundsdóttir, "Talking About a Gracious God," 238.

■ Conclusion

The experience of Steingrímsson and his people, seen through his theological lens, is an experience of God's grace and God's use of natural disasters to teach tormented human beings a lesson. When I compare Luther's understanding of who God is, according to his explanation of the first article of the Apostolic Creed, to the image of God in Steingrímsson's description of his experience of the tragic events at the end of the eighteenth century, there are both similarities and differences. What is striking about Steingrímsson's God, who is indeed good and caring, is how God expresses this love and care for unfaithful people. This is a God who is "the source of good and bad" and uses "the fire" for disciplinary purposes, but who at the same time protects "his people" (not all, but some!) from being destroyed by the fire. This God is not evil but is nevertheless able to do things that are nothing but evil in human eyes, in order to bring about something good. This is what the pastor discerns when he looks back and evaluates the hardship he and his people have been through. Not everybody survived, but because of God's mercy and love, some did, including the pastor, who himself did everything he could to save his people from the cruel and unforgiving fire.

Living in a place like Iceland, which is rich in natural resources and beautiful nature, but where natural disasters have happened and continue to happen quite frequently, raises important questions not only regarding the relationship between God and our place but also about God. Steingrímsson's account is an attempt to respond to some of those questions. His story is one version of the story of our place, as a creation of God, and God as the creator of our place. I see another version of this story in what happened to the people who had to escape from their place overnight, because their mountain started spewing rocks, ashes, and deadly gases. This is also true to the story of those who experienced deadly avalanches hitting their villages. By coming to church, people expressed their belief that the church was a space where they could find comfort and compassion, where they could be themselves and say what was on their minds. In line with Job, people did not come to confess their sins but to cry out for help and ask difficult questions about God. Those stories are very different in many ways, yet they are all about people who live on this volcanic island, responding to events that threaten the future of their place, as well as their own future. Those stories are from different eras, when living conditions could not be more different. What these people share is their faith in God, the Creator of their place, and their prayers for help when they felt their place and their Creator had failed them.

■ Bibliography

Bernhardsdóttir Ásthildur Elva. "Learning from Past Experiences: The 1995 Avalanches in Iceland." *The Swedish National Defence College*, January 2001. https://www.researchgate.net/publication/320087744_Learning_from_Past_Experiences_The_1995_Avalanches_in_Iceland.

Billman, Kathleen D., and Daniel L. Migliore. *Rachel's Cry: Prayer of Lament and Rebirth of Hope*. Cleveland: United Church, 1999.

Brueggemann, Walter, and Tod Linafelt. *An Introduction to the Old Testament: The Canon and Christian Imagination*. 3rd ed. Louisville: Westminster John Knox, 2020.

Chirongoma, Sophia, and Ezra Chitando. "What Did We Do to Our Mountain? African Eco-Feminist and Indigenous Responses to Cyclone Idai in Chimanimani and Chipinge Districts, Zimbabwe." *African Journal of Gender and Religion* 27:1 (2021), 65–90. https://doi.org/10.14426/ajgr.v27i1.91

Eiríksdóttir, Sóley Eiríksdóttir, and Helga Guðrún Johnson. *Nóttin sem öllu breytti*. Snjóflóðið á Flateyri. Reykjavík: JPV, 2016.

Farley, Wendy. *Tragic Vision and Divine Compassion. A Contemporary Theodicy*. Louisville: Westminster John Knox, 1990.

Fjeldsted, Egill St. *Þrekvirki. Snjóflóðin í Súðavík og Reykhólasveit*. Reykjavík: Egill St. Fjeldsted, 2021.

Guðmundsdóttir, Arnfríður. "The Fire Alarm is Off: A Feminist Theological Reflection on Sin, Climate Change and the Melting Glaciers of Iceland and the Far North." In *Planetary Solidarity. Global Women's Voices on Christian Doctrine and Climate Justice*, edited by Grace Ji-Sun Kim and Hilda Koster, 135–54. Minneapolis: Fortress, 2017.

———. *Meeting God on the Cross: Christ, the Cross, and the Feminist Critique*. New York: Oxford University Press, 2010.

———. "Talking About a Gracious God: Speaking of God out of Experience." *Dialog: A Journal of Theology* 54:3 (2015), 233–40. https://doi.org/10.1111/dial.12184

Hugason, Hjalti. "'Honum sje eilíft lof og æra fyrir strítt og blítt', Jón Steingrímsson og Skaftáreldar." *Studia Theologica Islandica* 41 (2015), 35–56.

Hugason, Hjalti. "Sjálfsvitund á barmi taugaáfalls: Sr. Jón Steingrímsson og þróun einstaklingsvitundar." In *Menntaspor: rit til heiðurs Lofti Guttormssyni sjötugum,* edited by Dóra S. Bjarnason et al., 225–39. Reykjavík: Sögufélag, 2008.

Jennings, Willie James. "Foreword." In *Glimpsing Resurrection. Cancer, Trauma, and Ministry*, edited by Deanna Thompson, ix–xiii. Louisville: Westminster John Knox, 2018.

Johnson, Elizabeth A. *She Who Is. The Mystery of God in Feminist Theological Discourse*. New York: Crossroad, 1992.

Jones, Serene. *Trauma and Grace. Theology in a Ruptured World*. Louisville: Westminster John Knox, 2009.

Kleeman, Katrin. "The Laki Fissure eruption, 1783-1784." In *Encyclopedia of the Environment*, January 14, 2020. https://www.encyclopedie-environnement.org/en/society/laki-fissure-eruption-1783-1784/.

Luther, Martin. "Heidelberg Disputation." In *Luther's Works. Career of the Reformer, Vol. 1*, edited by Harold J. Grimm, 37–70. Philadelphia: Fortress, 1957.

Luther, Martin. "The Freedom of a Christian." In *Luther's Works. Career of the Reformer, Vol. 1*, edited by Harold J. Grimm, 333–77. Philadelphia: Fortress, 1957.

Luther, Martin. "The Large Catechism." In *The Annotated Luther. Vol. 2 World and Faith*, edited by Kirsi I. Stjerna, 279–415. Minneapolis: Fortress, 2015.

McFague, Sallie. *The Body of God: An Ecological Theology*. Minneapolis: Fortress, 1993.

———. "Reimagining the Triune God for a Time of Global Climate Change." In *Planetary Solidarity. Global Women's Voices on Christian Doctrine and Climate Justice*, edited by Grace Ji-Sun Kim and Hilda Koster. Minneapolis: Fortress, 2017.

Moe-Lobeda, Cynthia D. *Resisting Structural Evil. Love as Ecological-Economic Vocation*. Minneapolis: Fortress, 2013.

O'Connor, Katheleen M. *Job. New Collegeville Bible Commentary*. Collegeville: Liturgical, 2012.

Rafnsson, Sveinbjörn. "Um eldritin." In *Skaftáreldar 1783–1784. Ritgerðir og heimildir*, edited by Gísli Ágúst Gunnlaugsson et al., 243–62. Reykjavík: Mál og menning, 1984.

Sigurðsson, Ólafur Ragnar. *Undir gjallregni. Frásögn lögreglumanns af gosinu í Eyjum 1973*. Reykjavík: Bókafélagið, 2022.

Sveinsson, Óttar. *Útkall. Flóttinn frá Heimaey*. Reykjavík: Útkall, 2008.

Steingrímsson, Jón. *Fires of the Earth. The Laki Eruption 1783–1784*. Translated by Keneva Kunz. Reykjavík: The Nordic Volcanological Institute and the University of Iceland Press, 1998.

———. *A Very Present Help in Trouble: The Autobiography of the Fire-Priest*. Translated by Matthew Fell. New York: Peter Lang, 2002.

Thompson, Deanna. *Glimpsing Resurrection. Cancer, Trauma, and Ministry*. Louisville: Westminster John Knox, 2018.

Place and Space: Gathering Wisdom from the Life Work of Proto-Ecowomanist Fannie Lou Hamer

Melanie L. Harris[1]

Introduction

Ecowomanism is an approach to environmental justice that features the theo-ethical and sociocultural analysis and environmental scientific contributions offered by women of African descent. Insisting that we deconstruct colonial frames that have historically shaped environmental thought, thereby attempting to silence the voices of women of color and people of color, ecowomanism seeks to recover these voices and the liberatory lenses they use to create environmental justice. It moves in a

---

1. Melanie Harris is the Director of Food, Health and Ecological Well-Being and a professor of Black Feminist and Womanist Theologies, jointly appointed with African American Studies and the School of Wake Forest Divinity at Wake Forest University. She is registered as a coresearcher at the University of the Western Cape for the project on "An Earthed Faith: Telling the Story Amid the 'Anthropocene'."

How to cite: Harris, ML 2024, "Place and Space: Gathering Wisdom from the Life Work of Proto-Ecowomanist Fannie Lou Hamer", in EM Conradie & WJ Jennings (eds.), *The Place of Story and the Story of Place*, in An Earthed Faith: Telling the Story amid the "Anthropocene", vol. 3, AOSIS Books, Cape Town, pp. 121–136. https://doi.org/10.4102/aosis.2024.BK355.07

constructive way toward finding climate justice solutions that honor the interconnectedness of all beings and understand the earth as sacred. By engaging the environmental justice paradigm, ecowomanist perspectives also highlight the importance of intersectional analysis as a tool for shaping new paths of climate justice that recognize the importance of the connection between social and earth justice. That is, ecowomanism connects social justice issues such as racial justice, gender justice, and sexual and reproductive justice to environmental justice. Thus, it claims the importance not just of all Black lives but also of all life on earth and earth itself as living beings worthy of climate justice. Regarding racial justice specifically, ecowomanist approaches suggest that the logic of domination embedded in white supremacy is similar to the logic of domination embedded in anthropocentric attitudes that value the earth and nature as an object rather than a living organism. It argues that earth and earthlings are a part of a living, breathing earth community.

■ Origins

Interdisciplinary in its scope, ecowomanism has many points of origin, including within the fields of African American environmental history, environmental science, sociology, theology, ethics, and religion. The first time the term ecowomanism appeared in a publication was in "Ecology is a Sistah's Issue too: The Politics of Emergent Afrocentric Ecowomanism" by Shamara Shantu Riley.[2] Womanist ethicist Chandra Taylor Smith is cited as one of the first scholars to conceptualize an analytical interdisciplinary conceptual frame that engages the Black women's literary tradition, womanist theology, and ecology together. Her dissertation, *Earth Blood and Earthling Existence: A Methodological Study of Black Women's Writings and Their Implications for A Womanist Ecological Theology*, was published by University Microfilms (UMI) Dissertation Services in 2001. Expanding upon the work of African American environmental historians looking into the contributions of Black male environmentalists such as George Washington Carver, Dianne D. Glave sought to highlight the connections between Black women's spiritual activism and liberating Black theologies preached from the pulpits of Black churches to move congregants to fight for environmental justice. In her essay, "Black Environmental Liberation Theology,"[3] she coins the term *Black environmental liberation theology* and makes special note of how Black liberation theology influenced Black women church leaders in Warren County, North Carolina, to protest the dumping of toxic waste in thirteen counties across the state in 1978. Although Glave does not use the term ecowomanism, she notably writes

---

2. See Riley, "Ecology Is a Sistah's Issue Too."

3. See Glave, "Black Environmental Liberation Theologies."

about the significant contributions that African American women made to the environmental justice movement by reminding readers it was their spiritual activism that gave birth to the environmental justice movement in the United States of America (USA).

Foregrounding the work, contributions, and thought of African and African American women to environmental thought is the primary goal of ecowomanism. This can be seen throughout my previously published work *Ecowomanism, Religion and Ecology*[4] and in *Ecowomanism: African American Women and Earth Honoring Faiths*.[5] These foundational texts are central to environmental studies in that they explicate the ecowomanist method and model intersectional analysis, key to any approach to climate justice within the larger field of environmental studies. The first book illustrates the global network of climate justice work being led by women of African descent and the decolonizing lens these scholars use to approach climate issues. This book names three hallmarks of ecowomanism: (a) the importance of intersectional methodology detailing how race, class, gender disparities, and other social categories of difference interlock to construct modes of environmental injustice; (b) adopting a global scope when facing climate issues—ecowomanism does this by recognizing global links in the African and Asian diasporas around the environmental crisis; and (c) an imperative for earth justice, thus opening the theoretical door to understanding earth-honoring faiths and moral and ethical frameworks influenced by African, Native American, and Indigenous cosmologies, religious rites, and cultures. Notably, the work of ecowomanist scholar Sofia Betancourt, author of *Ecowomanism at the Panama Canal: Black Women, Labor and Environmental Ethics*,[6] builds upon each of these hallmarks and expands ecowomanist analysis to include attention to a fourth hallmark, the interreligious nature of ecowomanist thought and its contributions to the discourse of religion and ecology.

It is well known that womanist theologians and ethicists Delores S. Williams, Katie G. Cannon, and Jacquelyn Grant established womanist theology from a Christian frame, borrowing the term womanist from Pulitzer Prize–winning author and pioneering Black literary scholar and poet Alice Walker. However, the scholarship of third and fourth waves of womanist theologians, ethicists, and religious scholars expanded the discourse beyond Christianity to include a variety of religious, spiritual, and Indigenous traditions that honor Black women's moral codes, their stance as leaders in their communities, and the impact of their religious perspectives

---

4. See Harris, *Ecowomanism, Religion and Ecology*.

5. See Harris, *Ecowomanism: African American Women and Earth Honoring Faiths*.

6. See Betancourt, *Ecowomanism at the Panama Canal*.

on religion, society, and planetary care. One such leader, highlighted as a model of ethical living by womanist scholars, including Rosetta Ross and Karen Crozier, is Fannie Lou Hamer.

It is an inquiry into her life, activism, and theo-ethic of love that presents her as a proto-ecowomanist and opens our theological imaginations to approach the following questions: What difference does it make to the story of God's love to describe it in the geographical terms of place and space, as understood through African cosmology and the theo-ethic of Fannie Lou Hamer? How do her acts of resistance against white supremacy and economic injustice through the creation of the Freedom Farms in the late 1960s and early 1970s establish an important model for environmental justice as evidence of God's love?

■ African Cosmology and African American Stories of Place and Space

The Creation by James Weldon Johnson—excerpts:

> And God stepped out on space,
> And he looked around and said:
> I'm lonely—
> I'll make me a world.

> And far as the eye of God could see
> Darkness covered everything,
> Blacker than a hundred midnights
> Down in a cypress swamp.

> Then God smiled,
> And the light broke,
> And the darkness rolled up on one side,
> And the light stood shining on the other,
> And God said: That's good!

> Then God reached out and took the light in his hands,
> And God rolled the light around in his hands
> Until he made the sun;
> And he set that sun a-blazing in the heavens.
> And the light that was left from making the sun
> God gathered it up in a shining ball
> And flung it against the darkness,
> Spangling the night with the moon and stars.
> Then down between
> The darkness and the light
> He hurled the world;
> And God said: That's good![7]

---

7. See Johnson, "The Creation," 15–20.

The Creation, a poem by James Weldon Johnson, written in 1927, is a classic example in African American literature showcasing the brilliant creativity and literary genius of African American poets, writers, and scholars of the nineteenth century. The poem weaves theological depth with a spiritual power that charts the path for multiple generations of people of African descent to cast their lot for freedom and consistently fight for justice in the USA and around the world. It also paints a new picture of God. Specifically, the poem invites reflection on a new hermeneutic of the divine Creator as having both fatherly and motherly characteristics. Womanist and literary scholars throughout the discourses of theology, religion, and literature often reflect upon the image of God as a mother, "a mammy bending over her baby," captured in Johnson's poem. They suggest that this introduces a hermeneutic in Black religion that connects the origin story of the cosmos with the presence of the feminine divine, thus shattering any normative gender claim that God is solely male. Even beyond the act of defying gender norms, scholars also read the *Creation* poem text as presenting the African American voice (poet) as narrator, griot, storyteller, and the voice who names God. This role, explained best by womanist theologian Delores S. Williams,[8] who highlights Hagar in the biblical story as the first woman, and first Black woman, to name God, centralizes the African American voice, hermeneutic, and perspective in the doing of theology.

That is, similar to the mode of resistance in most African American literature that decenters white norms and shakes the notion of whiteness from countless epistemologies, Johnson's *Creation* visibilizes Black thought and Black theology and opens wide the discourse of Black epistemologies. Preceding the literary painting of new ecological worlds in the writings of Alice Walker ("Temple of My Familiar"), Octavia Butler ("Parable of the Sower"), or Afro-futurist writers N.K. Jemisin or Nnedi Okorafor, Johnson's *Creation* sermon sets the stage as an example in African American literature for readers to consider a cosmos outside of a Eurocentric image of God as creator, white, and male.

Even beyond the inclusion of multiple genders as God, the interruption of the whiteness of God, thus shaking the alignment between divinity and whiteness, womanists' readings of *The Creation* also note a deep interrelatedness between all beings and the sacredness of all beings as reflected in African cosmology. Religious theorists familiar with African cosmologies will remark upon these two points in Johnson's *Creation*, noting firstly the multigendered, nondual and creative nature of God and, secondly, the inherent interconnectedness of all of creation, reflected in how each element of nature is set into motion only in being connected to another part of creation. In this sense, then, religionists, theologians, and

---

8. See Williams, *Sisters in the Wilderness*.

scholars investigating origin stories must consider the interconnectedness of all beings as central to the framing of the cosmos. They must note the "falling away" of dualist categories, such as gendered categories, when describing God. Using a postcolonial lens or hermeneutic to read Johnson's poem in the present day, one can also glean from the poem that to witness the divine at work is to see evidence of nonduality and understand that this divinity weaves through (and is) all creation, making all beings interconnected. Two tenets that frame the African cosmology illustrated in James Weldon Johnson's *Creation* then highlight the interconnection of all beings and the sacredness of all beings.

Outlining the debates within Black religious discourse that argue whether or not remnants of African religious and cultural traditions show up in African American religious life after and through the transatlantic slave trade goes beyond the scope of this essay. However, it is important to note that ecowomanist approaches accept the argument that African retentions and "remnants," in the words of womanist and religious historian Rachel E. Harding, are sewn throughout African American spirituality. In a book by Rosemarie Freenie Harding, coauthored with her daughter Rachel E. Harding, *Remnants: A Memoir of Spirit, Activism and Mothering*, Freenie Harding discusses how the remnants of African spirituality show up in Black religion and reflects on how Black peoples' persistence in life, despite their experiences of racial trauma in the USA, shapes their religion. Rosemarie Freenie Harding writes:

> I have learned from the work of historian of religions Charles Long and dramaticist-philosopher George Bass. The meaning of religion for Black folks, they insist, is in the heart of our history, our trauma and our hope. It is what makes us indigenous to this place, to modernity. As Long puts it, Black religion is the way we have oriented ourselves—over the centuries in these Americas and extending back before our arrival on these shores to—"mash out a meaning" of life in the midst of tremendous suffering and pain. Religion, in this sense, is not simply a doctrine of faith or the methods and practice of church; rather it is all the ways we remind ourselves of who we really are in spite of who the temporal powers say we are. Religion, is how we situate ourselves, how we understand ourselves, in a particular place and time vis-à-vis Ultimate reality, vis-à-vis God [...] It is how we make sense and joy out of our human experience.[9]

Considering that the remnant itself is an act of joy and resistance for Black peoples, the remnant itself becomes an element and reflection of the African cosmos. That is, whether story, poem, sacred song, or art song, from the African American tradition of spirituals, the secular beat of rhythm and blues of Aretha Franklin, or the life-affirming beats of hip-hop, the remnant of African cosmology lives throughout Black life. It lives through

---

9. Rachel E. Harding and Rosemarie Freeney Harding, *Remnants*, 117.

and beyond racism, classism, sexism, and homophobia; it lives in, through, and beyond the joyful triumph of true justice; it cries out in the unifying spirit of protest, insisting that Black lives matter. The remnant itself is evidence of a belief in a cosmos in which all beings are connected, and this brings meaning and life to Black and Indigenous people everywhere. The remnant, then, is a reminder of life energy, *Ashe* itself.

For ecowomanism, this is the work of God's creative love. God's love is wide enough and deep enough and the only thing that is all-pervasive enough to go through all history and permeate and transform even the cruelest realities of human suffering in the cargo holds of slave ships as African peoples were stripped of their royalty, identity, and personhood to be forced into chattel slavery. Only God's love is deep enough to be present to the enslaved African there, in that moment then, and also be present in this moment now when Black peoples across the earth are still crying out for solidarity in the work of creating and sustaining earth justice. Only God's love is rich enough to renew and transform the soil, overrun by consumerism and greed, as evidenced in the white supremacist practice of sharecropping in Mississippi and in many places and spaces throughout the South that African peoples called home.

■ Fannie Lou Hamer: Place and Space

> I never thought of leavin' the South. I love it. I was born here. I watched my folks take a double-blade ax and chop down trees like that pecan there. A lot of this land the white folks is usin' now, I watched my parents clean it up. We're not trying to take this land away from them, but we got a right to stay here.[10]

> This may not seem like much to other Americans who constantly move about the country with nothing but restlessness and greed to prod them, but to the Southern black person brought up expecting to be run away from home—because of lack of jobs, money, power and respect—it was a notion that took root in willing soil. We would fight to stay where we were born and raised and destroy the forces that sought to disinherit us. We would proceed with the revolution from our own homes.[11]

These two quotes, the first by Fannie Lou Hamer and the second by Alice Walker, illustrate the meaning and importance of place and space for Black Southerners, especially those working in the civil rights movement. Hamer, a civil rights (and later human and environmental rights) activist, became well known as a leader in the civil rights movement for her audacious courage and plain-spoken oratorical genius at speaking truth to power.

---

10. Fannie Lou Hamer, cited in Egerton, "Fannie Lou Hamer," 106.

11. Alice Walker, "Choosing to Stay at Home," 161.

Hamer is best known for her role as co-founder of the Mississippi Freedom Democratic Party and the testimony she gave at the Democratic National Convention in 1964 about the pervasiveness of racial injustice throughout the South. While considerable effort was made to silence her testimony by President Lyndon Johnson himself, her act of speaking truth to power was later televised, detailing horrific stories of racial violence in the South. Arguing for the state delegation to be integrated, Hamer fought for justice on the center stage of the United States (US) government. Later that year (1964), Hamer helped to organize Freedom Summer and later helped to establish the National Women's Political Caucus.

■ Bound with the Earth: Fannie Lou Hamer Biographic Notes and Hamer's Courageous Contributions to Climate Justice

Born on October 6, 1917, Fannie Lou Townsend Hamer was born in the Mississippi Delta, the twentieth child of sharecropper parents, Lou Ella and James Townsend. Fannie Lou was especially close with her mother, Lou Ella, and reportedly gleaned most of her wisdom, know-how, and song from her mother, whom she cared for during her last years of life. However, Fannie Lou also learned a great deal of the art of storytelling and gained some of her oratorical skills and sense of determination from her father, James, a preacher in the community. As a child of sharecroppers, Fannie Lou recalls that she began picking 30 pounds of cotton a week at age six. In "To Praise our Bridges" in *Mississippi Writers: Reflections on Childhood and Youth*, she discusses the trickery and deceit experienced by many poor Mississippian youth, white and Black, who were forced to endure a life of exploitation through the system known as extended slavery: sharecropping.[12]

Karen D. Crozier's *Fannie Lou Hamer's Revolutionary Practical Theology: Racial and Environmental Justice Concerns*[13] is an extraordinary volume describing the many contributions that Fannie Lou Hamer made to the environmental justice movement and to practical theology. It illustrates the justice lens through which Hamer saw the world and points to community organizing methods that she used to raise awareness about the systemic nature of the intersecting oppressions, ecological and racial injustice, as both were sewn into the sharecropping system. Crozier's account follows Hamer's activism against sharecropping and poverty and her later protest work. Most importantly, the volume maps the arcs and movements in Hamer's thought, mines her spiritual wisdom, and documents her life work

---

12. Fannie Lou Hamer, "To Praise Our Bridges," 321.

13. Crozier, *Fannie Lou Hamer's Revolutionary Practical Theology*.

as a civil (and later human rights and environmental) activist. The book is also a practical theological account of Hamer's Christian theology and explores her religious and civil practice to reimagine theological themes that engage issues of racial, gender, economic, reproductive, and environmental justice. Noting the important identity of Fannie Lou Hamer captured in history books about her strength as a voting rights and civil rights activist, Crozier argues rightly that it was her training in community organizing, her leadership in organizations like the Student Nonviolent Coordinating Committee (SNCC) led by Ella Baker, and her firm Christian faith that emboldened her extraordinary courage to give her a more wholistic vision of justice. This vision is helpful for ecowomanism because it deconstructs hierarchies that divide humans from the earth and interrupts patterns of domination within anthropocentrism. Crozier writes:

> Hamer is instructive for creation care advocates [... *in that she*] argued that a right anthropology must be adopted to care for all of creation instead of operating on the assumption that everyone privileges human beings over non-human life forms in creation or the environment. Through Hamer's life as a sharecropper and only two generations removed from the institution of slavery, anthropocentrism is deconstructed, and the victims of colonialism, environmental violence, and global warming are given voice as humans who have intimate knowledge about and care for both humans and non-humans in particular spaces and places.[14]

Hamer's care for humans and nonhumans is rooted in a theo-ethic of love. From her Christian perspective and African roots, it is clear that she understood God as Creator and believed that every living being has sacred worth and value as a part of creation. Embracing an African cosmological vision, Hamer is a proto-ecowomanist whose life work reflects at least two primary tenets for ecowomanism: an embrace of African cosmology and an acceptance that human and nonhuman beings are sacred beings worthy of care on the planet.

Further study of Hamer's work by environmental justice scholars Monica M. White and Christopher Carter supports Hamer's significant place in the field as a proto-ecowomanist because she serves as a model of liberation and resistance in African American environmental history. Both scholars ground their arguments in a larger call for new and different approaches in environmental studies that showcase the agricultural knowledge of African Americans as modes of resistance. Rather than using a nonliberative frame that "emphasizes the conditions of exploitation [of African Americans] that primarily structured the relationship to the land—from slavery through sharecropping and Jim Crow era discrimination against black farmers,"[15]

---

14. Crozier, *Fannie Lou Hamer's Revolutionary Practical Theology*, 8–9.

15. White, "A Pig and a Garden," 33.

these authors create a liberative frame that highlights African Americans' knowledge and use of agriculture "as a basis for resistance," thus providing a frame for liberative ecowomanist ethics.[16] White explains:

> Contrary to perspectives that emphasize the legacies of oppression of black farmers and farmworkers in agricultural communities, and the long history of disenfranchisement/enslavement tied to agriculture, Hamer made food and its production an act of resistance and a strategy to build a sustainable community. Freedom Farms represented her vision of the centrality of food and agriculture in building self-reliant communities as a base for political activism.[17]

A careful look at White's analysis of Hamer's life and the Freedom Farm uncovers values maintained and modeled by Hamer and those who worked alongside her. These values undergirded these "self-reliant communities" and can be helpful in shaping an ecowomanist ethics of sufficiency. Having all but given up on the "American promise" and having toiled endlessly as a woman leader in the civil rights movement, combating sexism embedded within the movement to bring about change and true democracy, toward the end of her life, Hamer used her platform and international acclaim to fight for environmental justice. She worked to build up her community in a fashion that interrupted white supremacy, patriarchy, and poverty at the same time. Creating a model of resistance based on the right to and value of quality of life, Hamer's Freedom Farms Cooperative sought to "develop a cooperative intentional community, with housing, employment, educational opportunities, health care and access to healthy food, [reflective] of a self-determined, politically engaged, liberated community."[18]

As a proto-ecowomanist, Fannie Lou Hamer stands as a pioneering figure in the tradition. Not only is she a model of one who aligned the political, sociopolitical, religious–ethical, and racial–gender justice agenda with an earth justice agenda, but she also did so using her "organic intellectualism," her spirituality, and strategies in grassroots movement-building and organizing. The combination of sources from which Hamer found courage and fortitude to do her work present a wholistic model not only of an authentic life and noble character but also of a wholistic model of what it means to be a woman, womanist, spiritual being, leader, farmer, mother, partner, politician, and activist today. The model of leadership that Hamer provides is worthy of study, not only for the ways it can inspire activism in the present day, but also because of the fresh perspective on love and ethical ways of being with the earth that her work embodies.

---

16. White, "A Pig and a Garden," 22.

17. White, "A Pig and a Garden," 33.

18. White, "A Pig and a Garden," 33.

■ Hamer's Theo-Ethic of Love: Three Movements

Hamer's work in the civil rights movement shifted in the late 1960s as she became more aware of the inadequacies of civil rights activism. Attempting to align her Christian belief in love and her theological understanding of what mujerista theological ethicist Ada Maria Isasi-Diaz called the "kin-dom" or family of God with her activism, Hamer developed a three-stage theo-ethic of love, namely focusing on the practice of Black self-love, standing up for principles of fairness and equity, and communal freedom inclusive of the earth.

The first stage begins with a focus on self-love and self-respect and can be understood as love lived out loud but strengthened from within. According to theologian Karen D. Crozier, the first phase ethic begins with the steps of self-reflection, self-awareness, and acceptance of the self as deeply loved and held in divine love and care. In this sense, self-love and self-respect are almost byproducts of self-acceptance, accepting that you are truly loved and worthy of love because you are created by the divine.[19] Regardless of the levels of racial hatred that Hamer and other Black Mississippians faced in life and in the work of racial justice in the South, Hamer insisted that Black people maintain a dignity, self-worth, and self-love that would protect them from becoming consumed with hate. As an activist, she insisted that the focus of racial justice work was not on "hating white people" but rather on loving and teaching Black people to love themselves. Crozier explains:

> Hamer was proud of the love she had for herself and all of God's human creatures. Loving her white enemy and oppressor was neither embarrassing nor hidden. Furthermore, it included self-respect, black pride, and standing up for beliefs even if they differed from one's white enemy and oppressor. The first-generation sharecropping parents of Mississippi transmitted a power and practice that protected Hamer and many other black children, from bitterness, self-destructive rage, hatred of white people, and violent retaliation toward white people.[20]

Hamer believed in human transformation. She held onto hope that eventually, this model of accepting God's love of the self and embracing self-love would transform the individual minds and hearts of white people throughout the South as well. She believed in the power of prayer. When we consider the importance of the contemplative practice of prayer as a self-care practice for activists in the work of racial justice, looking at Hamer

---

19. I am deeply grateful for the scholarship of Willie Jennings, especially for his description and theological and constructive analysis of *imago Dei* in *The Christian Imagination*.

20. Crozier, *Fannie Lou Hamer's Revolutionary Practical Theology*, 19.

is instructive. As Crozier explains it, a look into Hamer's contemplative practice of prayer sheds light on how human and social transformation were connected for her. Crozier writes that Hamer:

> [S]eemed to have nurtured an interior life that enabled her to publicly resist her white oppressors while remaining open to embrace her white sisters and brothers as she loved and respected herself.[21]

Noting that Hamer's theo-ethic embraced the private and public spheres, Crozier goes on to explain the second and third stages of Hamer's theo-ethic as giving attention to the systemic intersecting realities of white supremacy, political disenfranchisement, and racial injustice. Beyond a framework of civil rights that seemed to shift its revolutionary stance when under threat of losing funding and financial support from the white middle-class, Hamer's insistence on racial reparations and equity was steady throughout her leadership.

Moving beyond the interior frame of a love ethic reflecting God's love of the self, self-love and respect, and the ways that accepting this form of love can transform one's interior life to provide strength for spiritual steadiness necessary in the work of justice, Hamer also recognized that her freedom and liberation were tied up with the freedom and liberation of all whose rights were being cut off by the logic of white supremacy, gender, and economic inequality. Attending to the systemic ways in which racism, sexism, and economic injustice ravaged the lives of sharecropping families in Mississippi, further dehumanizing them, in the second phase of Hamer's theo-ethic she used her speeches to constantly illustrate how profound the gap was between justice and the normative practice of economic exploitation, racial violence, and white supremacy. Hamer's insistence that the dignity and respect of all be honored deepened the call for justice.

The third phase of Hamer's theo-ethic grows out of the first and second phases and places attention on the cultivation of communal freedom, inclusive of the earth community. Similar to how the arc of an ecowomanist intersectional analysis lens expands race, class, and gender analysis to be more inclusive of earth justice and the intersecting realities of oppression facing the earth, Black and Indigenous people, and women of color, the third phase of Hamer's theo-ethic notes a shift in her thought to expand her approach for justice to be inclusive of the earth community. Hamer's approach shifted in 1963. In June of that year, Hamer was severely beaten by white prison guards in a Winona jail. Upon reflecting on her experience as a survivor of brutal racial and gender violence, she became convinced that the protective layer of self-love, which coated her heart with compassion

---

21. Crozier, *Fannie Lou Hamer's Revolutionary Practical Theology*, 122–23.

so that she did not fall prey to a sense of victimization or hate, grew from her theology, faith, and sense of communal love. This love that came from the people in her community and the land helped her heal from the trauma(s) she experienced during her life. The beauty of the land and the steadiness of pace and place so often described in poetry of Black southern writers gave her a deep sense of belonging. For Hamer, the healing love and support she received from people and place helped her reclaim herself, her passion for life, and her calling to do justice. By looking through an ethical lens, we can note that Hamer's third phase of a theo-ethic presents an approach to living the "more excellent way" of love. As we have discussed, Hamer's faith and theology allowed her to process the pain of the abuse she suffered while also acknowledging a deep sense of love and place that surrounded her. This acknowledgment of the truth and power of love allowed Hamer to engage in a true process of human transformation. Rather than becoming caught in feelings of hatred or revenge, Hamer chose to focus on the life-giving quality of place and community as a sacred energy that was greater than any oppression. This love allowed her to heal and reclaim her own freedom.

It is important to note that freedom of mind, heart, spirit, and body were not just for her. Hamer strongly believed that "nobody is free, until everybody is free"[22] (a central claim of Black feminism). Thus, Hamer's sense of freedom includes communal freedom, including the earth.

The third phase of Hamer's theo-ethic can perhaps be better understood by reflecting on the concept of *taking the arrow out of your own heart*, as explained by Alice Walker. In her essay "Suffering Too Insignificant for the Majority to See," and her more recent book of poetry *Taking the Arrow Out of Your Own Heart*,[23] Walker explains that true healing, and thus freedom, stems from the wisdom that it is good to focus on one's own healing rather than to focus on seeking revenge. Healing the self and finding the courage to venture into all aspects of the self that need healing from violence and racial trauma, Walker suggests, is not just an individual journey but a collective journey of healing. Walker writes:

> Suppose someone shot you with an arrow, right in the heart. Would you spend your time screaming at the archer, or even trying to locate him? Or would you try to pull the arrow out of your heart? White racism, that is to say, envy, covetousness, and greed (incredible sloth and laziness in the case of enslaving

---

22. See Fannie Lou Hamer's speech, "Nobody's Free Until Everybody's Free," delivered at the founding of the National Women's Political Caucus, Washington, DC, July 10, 1971, last accessed May 12, 2023, https://academic.oup.com/mississippi-scholarship-online/book/29348/chapter-abstract/244099842?redirectedFrom=fulltext.

23. See Walker, "Suffering Too Insignificant for the Majority to See," and *Taking the Arrow Out of Your Own Heart*.

> others to work for you), is the arrow that has pierced our collective heart. For centuries we have tried to get the white archer even to notice where his arrow has landed; to connect himself, even for a moment, to what he has done. Maybe even to consider apologizing, which he hates to do. To make reparations, which he considers absurd.
>
> This teaching says: enough. Screaming at the archer is a sure way to remain attached to your suffering rather than easing or eliminating it. A better way is to learn, through meditation, through study and practice, a way to free yourself from the pain of being shot, no matter who the archer might be.[24]

Similar to the wisdom offered by Walker, Hamer's theo-ethic is a model based on the same wisdom. Her theo-ethic leans into the notion that the destructive anger and hatred that consumed the men who beat her had little power to control her life or rob her of a sense of agency to continue living and fighting for justice. Hamer's life, speeches, and activism illustrate that she insisted on exposing the truth that love has the power to transcend hate. She insisted on healing. She insisted on healing herself, her community, and the land and doing so as an expression of compassion for every being, including the very ones who beat her. By choosing this radical form of love over hatred and developing a spiritual discipline of "praying for your enemies," Hamer engaged a deeper and more transformative power, enlivening the third phase of her theo-ethic. The act of "loving" one's enemy, including seeking to help, aid, and understand the social plight and circumstance of many white Mississippians living in poverty in her time, opens the frame of love from individual to communal. Rather than considering only how an individual is impacted by trauma, racial hatred, and violence, Hamer presents a mode of compassion that invites a practice of humanizing "the enemy" and acknowledging that the enemy, too, is a part of the earth, with sacred worth and value. Instead of leaning into individualism and dualistic thinking, Hamer's third phase of the love theo-ethic presents love as a root for human and social transformation that helps us see our connectedness to the earth and to each other.

For Hamer, this sense of connectedness extended beyond humans. Hamer also shared a compassion for the earth, and a love for nature, reflective of an African cosmology. She recognized that the land, the earth, was also being co-opted in systems of white supremacy and economic exploitation like sharecropping. The third phase of her theo-ethic widens the scope of communal freedom, not only to include the "enemy" as fellow earthlings but also all of earth as a part of the community.

---

24. Walker, "Suffering Too Insignificant for the Majority to See," 105.

■ Conclusion

As a proto-ecowomanist, Fannie Lou Hamer's life of faith-inspired activism provides a model of God's love in the world. When we consider the difference it makes to the story of God's love in terms of the place that Hamer called home and the earth community that she considered full of sacred worth and value, we understand more deeply the interconnectedness of African cosmology. To understand all things as sacred interrupts a logic of domination and therefore also dismantles the logic of white supremacy. Hamer's theo-ethic suggests that our theological imagination leans into understandings of God's love when God's love is experienced as social and earth justice. Contrary to dualistic patterns, even within Christianity, that separate the earth from heavenly or divine realms, Hamer's theology suggests that God's love manifests in a deep sense of community and place. Through her work as a civil rights, voting rights, and environmental justice activist, we see how her ecowomanist intersectional analysis informed her sense of justice and pioneering work as the founder of the Freedom Farms.

Creating spaces where all beings can flourish was a part of the work of God's love for Fannie Lou Hamer, and the tools she used should not be overlooked for doing justice today. Through analysis of her values, we observe that her contemplative life of prayer, her singing, and her life as a farmer, deeply connected with the earth, profoundly informed her leadership. Her activist strategies rooted in community organizing and the ethical frame shaped by her Christian theology rooted in Black self-love, self-respect and honoring all of creation provide us with a model and an approach for environmental justice work today. Contrary to environmental studies approaches that negate the power of Hamer's activist work and contributions to environmental justice because she was a Black woman, in this essay I have argued that we may restore the importance of Hamer's model of environmental justice work as a model for ecowomanism. By reclaiming Hamer as an ecowomanist mother or proto-ecowomanist, we can see evidence of God's love earthed. This is Hamer's offering, and we are invited to live and breathe new life into these values and share in the belief that love will always extinguish hate.

■ Bibliography

Betancourt, Sofia. *Ecowomanism at the Panama Canal: Black Women, Labor and Environmental Ethics.* Lanham: Lexington, 2022.

Crozier, Karen D. *Fannie Lou Hamer's Revolutionary Practical Theology: Racial and Environmental Justice Concerns.* Boston: Brill, 2021.

Egerton, John. "Fannie Lou Hamer." In *A Mind to Stay Here: Profiles from The South*, 92–106. London: Macmillan, 1970.

Glave, Dianne D. "Black Environmental Liberation Theologies." In *To Love the Wind and the Rain: African Americans and Environmental History*, edited by Diane Glave and Mark Stoll, 189–99. Pittsburgh: University of Pittsburgh Press, 2005.

Hamer, Fannie Lou. "To Praise Our Bridges." In *Mississippi Writers: Reflections on Childhood and Youth, Volume 11: Nonfiction*, edited by Dorothy Abbott, 321–30. Jackson: University Press of Mississippi, 1985.

Harding, Rachel E., and Rosemarie Freeney Harding. *Remnants: A Memoir of Spirit, Activism and Mothering.* Durham: Duke University Press, 2015.

Harris, Melanie L. *Ecowomanism: African American Women and Earth Honoring Faiths.* Maryknoll: Orbis, 2017.

———. *Ecowomanism, Religion and Ecology.* New York: Brill, 2016.

Jennings, Willie. *The Christian Imagination: Theology and the Origins of Race.* New Haven: Yale University Press, 2010.

Johnson, James Weldon. "The Creation." In *God's Trombones: Seven Negro Sermons in Verse*, edited by Henry Louis Gates, 15–20. New York: Penguin, 2008.

Riley, Shamara Shantu. "Ecology Is a Sistah's Issue Too: The Politics of Emergent Afrocentric Ecowomanism." In *Ecofeminism and the Sacred*, edited by Carol Adams, 191–204. New York: Continuum, 1993.

Walker, Alice. "Choosing the Stay at Home." In *In Search of Our Mothers' Gardens: Womanist Prose*, 158–70. New York: Harcourt Brace Jovanovich, 1983.

———. "Suffering Too Insignificant for the Majority to See." In *We Are the Ones We've Been Waiting For: Inner Light in a Time of Darkness Meditations*, 88–110. New York: The New Press, 2006.

———. *Taking the Arrow Out of Your Own Heart.* New York: Simon and Schuster, 2018.

White, Monica M. "'A Pig and a Garden': Fannie Lou Hamer and the Freedom Farm Cooperative." In *Food and Foodways* 25:1 (2017), 20–39. https://doi.org/10.1080/07409710.2017.1270647/

Williams, Delores S. *Sisters in the Wilderness: The Challenge of Womanist God Talk.* Maryknoll: Orbis, 1993.

An Arawete Cannibal Theology of Noncreation: Patriarchy, Capitalism, and Food

Eneida Jacobsen[1]

■ Placing My Essay

My essay is an exploration of Arawete creation theology. I rely on Eduardo Viveiros de Castro's anthropological work on the Arawete people,[2] whereas the Brazilian anthropologist uses the philosophy of Gilles Deleuze to describe an Amazonian ontology of becoming, I privilege Karl Marx's and Friedrich Engels's dialectical materialism in framing Arawete food-centered ontology. Because, having been gods, humans

---

1. Eneida Jacobsen, or Aeneid Jacob, is an adjunct professor of philosophy at Villanova University in Philadelphia. They are registered as a coresearcher at the University of the Western Cape for the project on "An Earthed Faith: Telling the Story amid the 'Anthropocene'."

2. I was first introduced to Eduardo Viveiros de Castro's work in Amerindian perspectivism in a graduate course titled "Figures of Critique in the Borderlands," which was taught by my current dissertation advisor Dr. Gabriel Rockhill in 2015. Since then, I have been interested in anthropology and the implications of its "ontological turn" for theology and philosophy. See especially Viveiros de Castro, *Arawete: Os Deuses Canibais*.

How to cite: Harris, ML 2024, "An Arawete Cannibal Theology of Noncreation: Patriarchy, Capitalism, and Food", in EM Conradie & WJ Jennings (eds.), *The Place of Story and the Story of Place*, in An Earthed Faith: Telling the Story amid the "Anthropocene", vol. 3, AOSIS Books, Cape Town, pp. 137–154. https://doi.org/10.4102/aosis.2024.BK355.08

An Arawete Cannibal Theology of Noncreation

have not been created, and because eating precedes creation, Arawete creation theology is a theology of noncreation. My essay is an interdisciplinary effort in the field of public theology. I extrapolate a people's notion of divinity to a critique of capitalist food production. I apply a see-think-change method of Christian liberation theology to an Arawete liberation theology.

The Arawete are a Tupi-Guarani people of farmers and hunters inhabiting the forests of the eastern Amazon jungle, near the Ipixuna River, a tributary to the Xingu River.[3] The Arawete are very religious, taking joy in talking about the gods [*Máí*] and singing and dancing to the songs [*maraká*] that the gods themselves take time and joy to create [*mara*] and sing through the mouths of a shaman [*peyo*].[4] Like many cultures, the Arawete teach a mythical story about the beginning of life on earth when a (masculine) god imposes cosmic order from or despite (feminine and human) chaos. Arawete creation narratives are stories about the domination of women by men and nature by humans through work. They narrate a story of class struggle to be overcome in heaven. Gods live in a place of no evils. Humans' re-creation will be their repositioning in the cosmic order as gods in a dancing, heavenly society, free from the necessity to work. In contrast to a concept of creation *ex nihilo*, many cultures seem to share the notion of divine creation as (masculine) transformation (Arawete), organization (Hebrew), and assassination (Babylonian) of already-existing matter. For the Arawete, whereas life was already there in the beginning, earth was not. Playing a rattle, masculine gods raised heaven away from earth and created animals from humans. Both masculine and feminine true gods participated in the process of resurrecting humans from their bones into gods after devouring the earthly flesh attached to the redeemable souls. In all cosmic spheres, eating preceded or was part of the process of creation of something that

---

3. The Arawete used to live near the Bacajá River. Because of the growing economic devastation of the Amazon region and consequent scarcity of food, and aggravated by disputes among tribes, the Arawete found themselves constantly fleeing from dangerous enemies as well as displacing weaker ones. In the 1960s, while fleeing from the Parakanã and dislocating the Asuriní, the Arawete approached the Xingu River, where they encountered hunters looking for feline fur, a practice declared illegal in 1967. The Arawete were relocated by Brazilian officials to the Ipixuna region in the 1970s. An estimated 36% of the Arawete population died during a difficult journey through the forest and from diseases caught from white people. By 1977, the first census conducted by the Brazilian Federal Agency for Indian Affairs (Fundação Nacional dos Povos Indígenas [FUNAI]) counted 120 Arawete individuals. That number doubled by the year 2000, when 278 people were surveyed. See Ribeiro, "Araweté: A Índia Vestida"; Viveiros de Castro, *Arawete: Os Deuses Canibais*, 130-35; Viveiros de Castro, Caux, and Heurich, *Araweté: Um Povo Tupi da Amazônia*.

4. The verb *peyo* means to "blow wind" or "make blow wind," which could suggest a correlation with the Christian notion of Spirit. A "pastor" here is a "spirit-blower."

did not yet exist. The absence of time is not something that could be conceived. All that exists happens as relational becomings.[5] The life relationship between eater and eaten is a fundamental category of Arawete thinking and social organization. To eat, and to be eaten, is the dialectic of all sociocosmic existence. To be is to struggle as classes placed within cannibal relations of becomings.

Arawete society and theology are, I am claiming, food-centered and patriarchal. Is this not a contradiction? A society can be both feeding and nonfeeding by feeding less to feeders. Being food-centered, Arawete theology is anticapitalist. To eat sustains the living. Being patriarchal, an Arawete hierarchical division of labor fits the elemental form of a protocapitalist society in which the eating (consuming) class holds power over the feeding class. In paradise, food used to be available without the intermediation of labor. There reigned equality. Like Eve and Pandora, the Arawete blame Tadïde for humanity's fall into a life that requires labor to subsist until dying. Misogyny, that is, social contempt for women, has been taught for millennia through many narratives of creation and fall. Why? I take the Arawete society as a case study comparable to global capitalism in its fundamental social structure, even though the division of labor is not as accentuated among the former. My Marxist-feminist hypothesis is that gendered relations of production are not natural but taught through work relations and their accompanying ideologies for the benefit of eaters. A channel for the teaching of oppression is the belief in something existentially as profound as a god, or gods. To perpetuate social oppression, patriarchal societies have produced patriarchal narratives of creation and fall. On the bright side, earth will not remain separated from heaven forever. There is hope beyond death. The Arawete teach that gender differentiation will still exist in heaven, although only for pleasure and not as division of labor, because work will no longer be necessary. The ideal, heavenly society eats food free from the burden of work while dancing, singing, getting drunk, and enjoying life together. Society will remain enslaving only so long as work is the required mediation between food and the need to eat.

---

5. In the case of humans, as Viveiros de Castro writes, "alterity and becoming emerge as the defining quality and process of 'being' human." See Viveiros de Castro, *Arawete: Os Deuses Canibais*, 12. Human beings are what they are not yet, namely, gods. The meaning of Arawete words is highly dependent on "contexts of contrast." Viveiros de Castro, *Arawete: Os Deuses Canibais*, 205. By becoming a different being, different eating relations are at the same time established. To be is also not to be for there is no being that is not a becoming. To be is to become food, eater, and feeder.

■ An Arawete Theology of Creation in Between Places

The Arawete believe that the universe is composed of four distinct layers or supports [*hipã*]: earth [*iwi*], the underworld [*iwi katï*], the place of the gods [*maipi*], and the red sky [*iwã pïdï*]. The great "time of division" [*iwawa me*] between heaven and the firmament took place when the god Aranãmí, feeling offended by an insult [*ikirã*] by his wife Tadïde, abandoned earth, determined to live far away from humans [*bïde*]. Aranãmí was the last god to enter heaven and does not eat humans, is not cannibalistic [*piri o*], and therefore does not resurrect people. The story says that Tadïde "threw away his footprints" [*ipipa mara hetï*]. It is unclear how "footprints" can be thrown away and why Tadïde's words were insulting to Aranãmí.[6] Together with his nephew Hehede'a, Aranãmí sang and smoked tobacco while playing a rattle [*aray*]. A phallic object in Arawete culture, the rattle represents both masculine domination and the divine power of creation.[7] The god's action caused a rocky layer of the firmament to rise to the high sky, generating a catastrophe. Without the stone layer that used to be its foundation, the firmament dissolved [*ikie*] under the waters. Some versions of the story found in Arawete religious hymns speak of a river, while others mention rain as the cause of the flooding.[8]

During the universal flooding, most beings were devoured by a monstrous piranha [*pako oco*] and a monstrous alligator [*yicira oco*]. Among humans, only two men and one woman survived by climbing onto a bacaba [*pïdowa'i*] tree, a fruiting palm native to the Amazon rainforest. The trio are the ancestors of all humans. Indeed, the Arawete child usually has several fathers. It is expected that a pregnant woman will have sex with different men whose semen will compose the child's substance. Some divine beings avoided the flood monsters by consolidating a separate layer underneath the firmament. A large river runs in the underworld. The river has islands on which live the Tarayo, gods frequently invoked by shamans to help kill evil spirits, the Ãñí. Most terrestrial spirits are dangerous, especially to women. Particularly threatening is Iwikatihã, the river spirit whose name is associated with the underworld, who desires and kills human women, who will no longer be able to enter heaven once captured. In order not to be kidnapped by Iwikatihã, whose name must not be pronounced, women must not go bathe in the river by themselves. Some gods in heaven also desire women sexually. Women occupy the

---

6. See Viveiros de Castro, *Arawete*, 184.

7. Viveiros de Castro, *Arawete*, 240–41: it is said that a vagina "breaks the rattle of shamanism" [*aray mo-pē hã*].

8. See Viveiros de Castro, *Arawete*, 185.

position of prey across cosmic spheres. Another reason why it would be dangerous for a woman to be a shaman who walks up to the place where the gods live is that, having lighter souls than men, women would not be able to return from heaven to earth. Was Tadïde's insult the cosmic reason for women's placement in oppression?

All cosmic layers are in communication there "where the earth enters" [*iwi yece pã we*] or "where the sky descends" [*iwã neyi pã we*].[9] "We are in the middle" [*bïde ipi-te re*].[10] The earth has a flat and circular shape which elevates on its borders until it reaches the sky. Each cosmic layer has its own stars and moon glowing on the reverse [*ikipe tï*] of the neighboring sphere. The same sun [*karahï*] shines on all layers of the universe. During the night on earth, the sun shines both on the heavenly spheres and the underworld. The earth's moon is finite. Each moon is a new [or "other": *amïte*] moon which, by copulating with human women in their dreams, causes them to painfully menstruate. Celestial layers are made of stone. So are the gods' houses, pots, bowls, and axes. For the gods, stone is as malleable as clay is for humans. The first celestial sphere, of which the sky visible to earthly beings is its reverse, is called *maipi*, which means "place of the gods." *Maipi* is a land of no evils—and that entails no work. The Arawete call it "place of good existence" [*teka katï*].[11] In *maipi*, or heaven, live several divine races who abandoned earth as soon as Aranãmï and his nephew raised heaven, including the *Máï hete* or *Máï oho*, the "real" or "big" gods associated with thunders and lightning. All gods take pleasure in eating, dancing, having sex, decorating their bodies, walking in the forest, and getting drunk together. Whereas all gods compose songs, only "true gods" participate in humans' resurrection by smoking shamanic tobacco and cannibalizing the dead. The Arawete aspire to become like their gods, living forever in sensual enjoyment.

Gods and humans "simply exist" [*ikate*] and have eaten fruits from all sorts of trees for as long as we can remember. Spirits also simply exist, and those that are plants have been providing other beings with nourishment. At the same time, it is said that gods became gods and humans became humans. Divinization and humanization took place during the primordial time of dispersion, "when gods divinized" [*e'e me Máï odi mo Máï*][12] as they abandoned humans. Nothing "is" until there is becoming. Most beings that occidentalized people call "animals" have been created [*mara mi re*] by *Nã-Mai,*' the jaguar-god, out of revenge against the monstrous jaguar

---

9. Viveiros de Castro, *Arawete*, 186.

10. Viveiros de Castro, *Arawete*, 184.

11. Viveiros de Castro, *Arawete*, 219.

12. Viveiros de Castro, *Arawete*, 214.

[*Ñã nowí"hã*] for the killing of the god's mother. Nã-Máï used the usual shamanic instruments of creation when bringing animals into existence: the rattle and tobacco.[13] There is no word for "animal" in Arawete. Some categories employed to differentiate among beings regarding their function to the Arawete are "to eat" [*do pi*], "to be pet" [*temimã ní*], and "that eat us."[14] There are a few general terms, such as fish [*pïdã*] and birds [*irã*]. Fish were transformed by gods from cultural objects made of plants. Animals used to be humans. When animals were created, many became food for each other and humans. To live is to eat, and to eat or be eaten is to become living. Exceptions to this, beings who have not been created, are the South American tortoise [*jabuti*], who is said to be "very old" [*imí*], and some species of fish and insects, who also "simply exist."

The idea of "place" is contained in the Arawete word for "creation," which goes well with the topic of the present volume, which asks for the place of creation and the creation of place. The verb *mara*, to create, in literal translation, means "to put" or "to place." To create is to place someone somewhere. "Creation" is distinct from "fabrication" [*mõñĩ* or *apa*], which is a simpler elaboration from a raw material. Cultural objects such as the rattle are "fabricated," whereas hymns are "placed" or "created" by the gods. "Creation," Viveiros de Castro writes, is a "position of being," and to create is to "place as existent."[15] Although creation entails the appearance of a being that did not previously exist fully as such, there is no conception of creation out of nothing for the Arawete: animals were created from humans; the sky and firmament used to be undifferentiated matter; and humans used to exist as gods until they were abandoned by the gods.

Before the great separation between sky and firmament, in the Arawete paradise, all beings lived in harmony together. Darwinian evolution theory teaches that all life is related. So does Arawete theology. In the Arawete primeval paradise, gods, humans, and animals did not yet exist as distinct kinds. Nor were the earth, heavens, and the underworld separate cosmic spheres. Neither were humans marked by the paradox of holding both a terrestrial soul and one destined to become divine. The Arawete paradise was a place of perfection: there was no death, no sickness, and no work. Eating and having sex were activities of pleasure free from labor and suffering. There was no need for human reproduction or labor pain, as humans had always existed. Nobody had any knowledge about the

---

13. See Viveiros de Castro, *Arawete*, 224.

14. See Viveiros de Castro, *Arawete*, 223. The description of humans and other animals according to their diet is common for the Arawete. See Viveiros de Castro, 220.

15. Viveiros de Castro, *Arawete*, 223: "pôr como existente."

cultivation of plants, because agricultural labor was unnecessary. Fruits were collected without effort and with the greatest joy from abundantly available fruit-bearing trees. In paradise, there was no fire, thus, no cooking – although a lack of work, in this case, may be seen as negative, as the Arawete enjoy cooked food.[16] Not yet separated from eternal communion with the divine, humans lived in "pure nature": they did not work, they only ate plants, they were not eaten, and they did not kill, as beings for-humans-to-eat did not yet exist.

The fall from a primeval paradise marked the beginning of work, suffering, and death. Abandoned by the heavenly gods, threatened by evil spirits, persecuted by human enemies, condemned to a life of work, humans experienced many adversities in this life. At the same time, human reality used to be united with heaven, sharing a primordial status of being sacred. Life separated from heaven was a paradox, insofar as good and evil were intertwined. Among the pleasures of life were food, music, and sex. The days, which had been very hot until the great cosmic separation, started to cool once the periodicity of day and night was established by the primordial owl, who also gave celestial fire to the gods. Humans learned how to master fire and could cook their food. Humans learned agriculture, starting to grow plants to eat and revealing that knowledge to the gods. For gods, food is made available by plants without the necessity of the intermediation of agricultural labor. In human society, organized labor became a condition for existence. In many societies, from ancient times to today, division of labor has been instituted as gender oppression. As theologies are the products of their societies, and patriarchal societies generate patriarchal religions, it should be no surprise that myths that tell the origin of work frequently serve to justify the current order based on the social oppression of women.

Even though it may be argued that men and women share different responsibilities within the Arawete economy, that the division of labor is fluid rather than strict,[17] so that classes in cooperation do not oppress each other, there is, if not fully developed, a latent slavery to be identified. Women's freedom is limited in many ways. Women do not walk naked as men do. Instead, women wear an outfit that restrains their movements. Women are not warriors. As such, no woman can avoid the gods' cannibalism during the process of resurrection. Only those who have already died by killing an enemy [*awí*] have a painless path into heaven. Women are not shamans. They do not sing in the name of the gods during sacred rituals.

---

16. An ambivalent valuation of a precultural state is recurrent among Tupi-Guarani peoples. See Viveiros de Castro, *Arawete*, 259.

17. See Viveiros de Castro, *Arawete*, 47.

Men, too, have been abandoned, and yet, through the social institutions of family, religion, and war, they stand closer to the gods and may impose a social order and ontology that privileges men as the property holders of children and women. During religious rituals, women roll and light tobacco for their husbands. Women also serve men the alcoholic beverage which women have produced by chewing for hours on corn. Producers and consumers appear here as opposite classes: the one feeds, the other consumes. Not only as feeder does woman stand opposite to man. Woman is also food. In Arawete language, the sexual act is referred to as the act of eating a vagina or a woman. A woman's body becomes men's property or creation in sex when men sculpt [*maya*] and fabricate [*mõñĩ*] the vagina by stretching the labia majora until the vulva is deformed.

On average, it seems fair to say that women serve more of their lifetime as a productive labor force. The claim gains more force when considering that hunting, a primarily masculine activity, is not considered labor by the Arawete but rather as a hobby. Biologically, it is believed, as it was once thought in European science, that only men generate children. Women are nothing but a bag [*hiro*] in reproduction.[18] The Arawete child, as a rule, has several fathers. Men as a class might have seen a woman's body as the place where men's genealogical footprints become mixed and, to some extent, lost. Was that the (sexual) offense? In a time when men and women might have been just as agile, men constituted a class by ruling over women's bodies by numbers rather than by individual force. Arawete men agreed to share the children of their domination equally. There was a patriarchal collectivism. Woman has no posterity. Biological capital is men's. Women's "insult" works as a cosmic reason justifying to women why they must be servile feeders of men and leave no posterity.

Most living beings of all cosmic spheres—heaven, earth, underworld—physically eat and may end up being eaten. This has been so in the past and present and will be so in the future. For humans, to live is to eat, and to die is to be eaten. There is no life without food. To eat precedes any divine creation. Time, perceptible in becomings, is nothing outside of the relational activity of eating. The Arawete refrain from metaphysical abstractions. Resurrection is a physical action that implies the eating of flesh. Upon dying, humans' spirits undergo a profound transformation. Firstly, the person's evil soul is left on earth while the good soul walks up to heaven. Because, as Tupi-Guarani peoples generally believe, heaven is situated on the heaven's "side where the sun comes out" [*karahi rodïhã tĩ*], to help souls find their path up to heaven, they bury their dead with their faces toward the east. Upon entering heaven, good spirits' bodies are cooked

---

18. See Viveiros de Castro, *Arawete*, 456.

and devoured by the true gods, who leave no meat attached to the soul's bones. Nã-Máï's brother, Tiwawĩ,[19] is the true god who resurrects people from their bones after being cooked by *Máï dari* [divine grandmothers] and devoured by *Máï hete*. Then, the souls undergo one last reviving bath in boiling water and forget all earthly memories. Life taking place on earth in finitude is a temporary condition. For now, labor is the activity that mediates the Arawete people and their need for food, and that division is gendered. Only in heaven will division of labor cease, because work will no longer be necessary for living. Arawete people's destiny is to become gods and experience life's good tastes, textures, smells, and sounds without ever getting sick or being oppressed.

■ Patriarchal Societies and Patriarchal Theologies

Creation and fall narratives reveal how people productively relate to each other in their societies. Binary gender distinctions play a crucial role in various cultural narratives about primeval times. Patriarchal societies have taught narratives of creation with the purpose of inculcating patriarchy in people's minds and habits of relating to each other and nature. Patriarchy takes place at an economic level as division of labor,[20] and it is sanctified by religious ideas. "Man" and "woman" are social constructs historically developed as classes of production in conflict. Through rape and labor expropriation, marriage has been an enslaving institution for women. Simone de Beauvoir argued that religions have been invented by men and for men to rule over women. "Religions forged by men reflect this will for domination: they found ammunition in the legends of Eve and Pandora. They have put philosophy and theology in their service."[21] The idea of the divine is powerful. Who holds power in society either already does or wants to hold power over the idea of the divine as well. Patriarchy is the reason that explains why many narratives of creation and fall of various cultures speak of women in a demeaning way. Woman is matter, Beauvoir writes, whereas man is the law of "creation." "The Other is passivity confronting activity, diversity breaking down unity, matter opposing form, disorder resisting order. Woman is thus doomed to Evil."[22] Religion has been a tool

---

19. Tiwawĩ is known as *"bïde cĩ mõñĩ hã"* [fabricator of our bones]. Viveiros de Castro, *Arawete*, 238.

20. In patriarchy, the distribution of "labor and its products" benefits the father who holds his wife and children as slaves. See Marx and Engels, *The German Ideology*, 52.

21. Beauvoir, *The Second Sex*, 31.

22. Beauvoir, *The Second Sex*, 114.

to teach women's inferiority. By naming gods and speaking on their behalf, men have perpetuated their social power.

In the Babylonian creation narrative, the divine son Marduk shapes the cosmos out of the tortured body of the primeval Mother of all, Tiamat, separating heaven and firmament. "Divine brethren (brothers) banded together" (1:21), the *Enuma Elish* reads, "roiling the vitals of Tiamat" (1:23). The disturbance of Tiamat's inner organs may suggest sexual assault.[23] Following family disagreements, Marduk was induced by his brothers to kill Tiamat. Marduk blew wind and threw an arrow that tore apart Tiamat's insides (4:95-102). Marduk used one part of Tiamat's captured body to frame the sky and the stars, and the other part to shape the firmament (4:137-141). Tiamat's head became a mountain, and out of her eyes still run the Tigris and Euphrates rivers (5:53-55). From clay and the blood of younger gods that he murdered, Marduk molded human beings (6.1-39). According to Rosemary Radford Ruether, the Babylonian poem of creation was redacted from earlier Sumerian stories during Babylon's first dynasty (nineteenth to sixteenth centuries BCE), a historical period when, in place of a system of land rental from priests, patriarchal private property became a solid social institution controlled by political and religious elites.[24]

> Dead matter, fashioned into artifacts, makes the cosmos the private possession of its "creators." Even though the new lords remember that they once were gestated out of the living body of the mother, they now stand astride her dead body and take possession of it as an object of ownership and control.[25]

The product of the labor of the serfs was increasingly expropriated by rulers using military power. New territories were annexed through theft and violence. The rising of an empire is the expansion of patriarchal dominion over both bodies and lands.

Tiamat echoes in Genesis's Elohist tradition as "the deep" [*t-əhôm*] sea over which hovered darkness "and the spirit" [*wərûah.*] of god, or gods [*ĕlōhîm*].[26] Psychoanalytically, the image could be the recollection of familiar and yet mysterious waters during a mammal's early life in the womb. The darkness is contrasted with the light brought about by divine creating language. Whereas in the Elohist tradition, both man and woman are created in God's image (Gen 1:27), in the Yahwist source, woman is created after man because God thought "it is not good that the man should be alone" (Gen 2:18) and that he should have a helper [`ēzer*]. Contrary to

---

23. See Vajskop, "Finding Patterns in the Chaos," 64.

24. See Ruether, *Gaia and God*, 17-18.

25. Ruether, *Gaia and God*, 18.

26. See *King James Bible Online*.

a natural state in which man is born out of woman, here woman is created out of the man from his rib. Adam has the power to name Eve (Gen 2:23). The man is said to leave his family to start a new (patriarchal) unit (Gen 2:24). As the story goes, instead of fulfilling her role as helper, Eve decided to disobey the only order pronounced by her Creator. She eats the fruit from the prohibited tree of knowledge of good and evil and convinces Adam to do the same (Gen 3). Because of Eve's sin, women are condemned to pain in childbirth, and men struggle to cultivate the soil (Gen 3:16). Genesis's genealogies only name men who descended from Adam (Gen 5:1), erasing the memory of the families' mothers. Eve's fall was often used in theology's history to justify women's oppression. Ambrose of Milan wrote: "Adam was led to sin by Eve and not Eve by Adam. It is right and just that he whom she led into sin, she shall receive as master."[27] To rule over a woman means to expropriate her of her will, body, and offspring.

In Greece, Pandora, daughter of the male god Zeus, opened the box of all evil. Hunger, disease, and death exist because of the world's first woman's curiosity. From Pandora, the poet Hesiod writes, "is descended the female sex, a great affliction to mortals."[28] Because of Pandora's inability to follow the simple order of not opening a wedding gift, countless troubles roam among humankind: "full of ills is the earth, and full the sea."[29] Beauvoir observes: "Mythology's goddesses are frivolous or capricious, and they all tremble before Jupiter; while Prometheus magnificently steals the fire from the sky, Pandora opens the box of catastrophes."[30] Despite woman's destructive cosmic power, ancient Greek mythology still holds that reality remains under the direction of the very powerful father. Why are feminine divine figures portrayed in a negative way if not to justify women's oppression in society? Girls learn from a young age that, because they pertain to this magical substance socially identified as femininity, their actions are more likely to be inconsequential. If women are incapable of controlling their actions, then men must be entitled to control them, and women must remain in economic, political, and religious subordination even when women's labor sustains a people's economy.

For many Tupi-Guarani peoples, before anything existed there was Yamandu, the silence that illuminates all, also known as Nanderuvuçu, or Nhamandu.[31] Yamandu had no knowledge until the appearance of the

---

27. Ambrose cited by Beauvoir, *The Second Sex*, 133.

28. Hesiod, *Theogony*, 590–91.

29. Hesiod, *Hesiod, Works and Days*, 100–01.

30. Beauvoir, *The Second Sex*, 350.

31. See Oliveira, "*Mito Tupi Guarani da Criação do Mundo*," 19–22.

thought of the infinite. Within the self's thought and time, Yamandu desired knowledge of infinity. Yamandu became various birds and flew within the vastness of that which has no end. Yamandu became the primeval owl, and there was day and night.[32] Yamandu became the first hummingbird and rapidly flew in all directions of the luminous infinity of the divine self. Finally, Yamandu became the original harpy eagle and gained whole knowledge of the extension of the divine from far above. Having looked within the luminous self, Yamandu uttered the first words: "Let there be worlds." There was language, the thought of worlds, and soon music. Yamandu sang, and the music became worlds. Here, as for the Arawete, creation is a musical event. The difference, which could help support the view that not all Tupi-Guarani peoples have been patriarchal,[33] is that the creator Yamandu is neither masculine nor feminine. Divine substance precedes and transcends any differentiations in time. Even though Yamandu's first child, the creator of worlds, was masculine, soon that son, Tupã, wedded the goddess Arasy, or Jaci. Together, the divine couple created everything there is on earth: rivers, seas, forests, and living beings. Despite significant elements of gender equality, the figure of a woman responsible for a regrettable separation between humans and the gods is also recurrent in Tupi-Guarani mythology. In the Apapocúava-Guarani tradition, the masculine creator Nanderuvuçuu left earth following the announcement of infidelity and disbelief by his terrestrial wife Nandecy that the corn Nanderuvuçú had just planted was ready for harvest.[34]

■ An Arawete Critique of Capitalism: From Patriarchy Back to Nature

Women have historically played a central role in meeting the demands of nourishment for the human species. Gestation, breastfeeding, food gathering, apiculture, agriculture, and cooking have been crucial activities for subsistence in many societies, performed by a majority of women. Among hunter-gatherer societies, plant food collected by women may comprise up to 80% of the group's diet.[35] The frequently masculine activity

---

32. A Guarani shamanic chant goes as follows: "The true Great Spirit, the first / Existed before the first winds / Where it was anchored in the void-night / Made owl producing silences / And made to spin the manifestations of / The self facing the night, dressed as space." Translated from Portuguese by the author and from the Guarani by Jecupé, *Tupã Tenondé*, 30.

33. Archeological evidence suggesting a nonhierarchical system of labor cooperation has also been found, "thus contradicting the principle of universality of masculine domination." See Jácome and Furquim, "Teorias de Gênero e Feminismos na Arqueologia Brasileira," 4.

34. See Nimuendaju Unkel, *As lendas da criação e destruição do mundo*, 48–49.

35. See Lee, "What Hunters Do for a Living," 33.

of hunting, although very prestigious in many societies, including the Arawete, is not a source of nourishment as reliable as the food provided by the frequently feminine work of gathering fruits, growing plants, and cooking. Although essential for mediation between humans' need for nourishment and the sources of nature, women's work does not hold a high social status in many societies. Among the Arawete, activities that hold a high social status happen to be activities performed by men: hunting large animals, connecting the cosmic spheres using shamanic power, and killing human enemies. People are ascribed hierarchical roles in society according to the sex they were born into. Why did society become so specific in ascribing different roles to the sexes in society, creating binary genders as the most fundamental classes of labor division in human relations? Because genders have been associated with biological sexes, such distinction of labor may appear natural rather than a social construct. If the oppressed accept their condition as natural, the oppressors may rule uncontested.

Gender as a matter of division of labor may appear more clearly in a society such as that of the Arawete, which Viveiros de Castro describes as characterized by a greater simplicity of societal rules and material resources within Tupi–Guarani standards[36] than might be obvious at first when seeking an understanding of today's global society, in which long sequences of numerous production activities are developed by innumerable people. Yet the intensification of the division of labor on a global scale only deepens the alienation of the whole working class, which has lost all connection with the fruits of its labor. Labor division by gender is even more essential for global capitalism than for Arawete society. Now, patriarchal oppression is extended to most humans, animals, and the whole of nature for the sake of increased profitability by economic elites. As different as various societies may be from one another, they all rely on social production for survival. For many societies around the world that division has had a gendered connotation. In Arawete society, women are the main producers of a highly appreciated drink called *cauim*. Men are the sole consumers. The reason why women chew on corn for hours and for several days to produce alcoholic *cauim* is not because they are recognized for their labor but because tradition has placed women into the economic position of slaves. During religious rituals, male shamans sing songs they say to be of divine origin. Ordinary men dance and get drunk. Women repeat the songs learned from men's shamanic songs in ordinary life. Women serve alcoholic *cauim* to the men. From nature to the mouths of men: the mediation is women's labor. To be sure, in Arawete society, many activities are commonly executed by people of both genders. Men work hard, too, and Arawete

---

36. See Viveiros de Castro, *Arawete*, 47.

society, to some extent, only offers a parallel to the acute class division found nowadays in capitalist societies.

A similar relationship between women as producers and men as consumers is found in global capitalism. Already during the first decades of industrial capitalism in England, from about 1760 to 1820, women made up the largest portion of the working force across many industries, particularly textiles.[37] Even when housework became the rule for women under the ideology of men as breadwinners, double shifts never ceased to be the situation for poor women in capitalist societies. Since the 1970s' New International Division of Labor, when industrial production was largely transferred to poor countries working to satisfy the needs of consumption in rich countries, women in poor countries began to suffer a double expropriation both as unpaid labor in the home and cheap wage labor in factories.[38] A highly gendered global workforce produces essential and luxurious goods it will never consume.[39] Women's labor is decisive in many industries such as garments and electronics, averaging 45% of the entire paid workforce of more than 80 countries.[40] Women's work is typically valued less and is thus cheaper than men's. In the home, women's work as cleaners, cooks, caregivers, and so on is not valued as a type of work that society ought to pay for, as it is judged natural for women to look after their families without social support.[41]

Capitalism's international division of labor being strongly gendered is structurally similar to the patriarchy within Arawete family, society, and religion. These institutions rely on each other, whereas gendered slavery exists in all of them. The difference is that gendered division of labor among the Arawete has not yet developed further into a speciesism that situates humans as masters over the whole of the natural world. As such, I believe that Arawete theology offers a critique of capitalism's alienation from nature. Although the Arawete do not properly value women's essential contribution to their economy and biological reproduction, they do

---

37. See Berg, "What Difference Did Women's Work Make to the Industrial Revolution?"

38. See Mies, *Patriarchy and Accumulation on a World Scale*.

39. Eduardo Galeano puts numbers on the world's inequality in the following example: "With petroleum, as with coffee or meat, rich countries profit more from the work of consuming it than do poor countries from the work of producing it. The ratio is ten to one; of the $11 that the derivatives of a barrel of petroleum sell for, countries exporting the world's most important raw material get a sum total of $1 from taxes and extraction costs. Countries in the developed zone, where the oil companies have their head offices, get the other $10, the sum total of their own taxes—eight times larger than those of the producing countries—and the costs and profits of transport, refining, processing, and distribution, monopolized by the big corporations." See Galeano, *Open Veins of Latin America*, 157.

40. See Fetterolf, "In Many Countries."

41. Federici, *Wages Against Housework*, 1: "They say it is love. We say it is unwaged labor."

acknowledge that nourishment is a fundamental activity of the living. The gods enjoy eating as much as humans do. Without food, there is no life. There is, thus, no creation that precedes the relationship between eater and eaten. Eating is the condition for humans' survival and resurrection alike. We eat to survive, and we are eaten to resurrect. Arawete materialism is dynamic. To be is to eat. In capitalism, the interspecies imbalance between eaters who eat more than they produce on average is extended to an interspecies imbalance in which eaters do not give back to nature as much as they take.[42] Whereas patriarchy is perhaps as ancient as human religion, only under capitalism did speciesism become so widespread that the reproduction of natural means of subsistence of most species is now under threat of annihilation.

■ Conclusion

Creation is cosmic placement. The Arawete fall was a displacement of earth from heaven. Humans have been placed in finitude and suffering. On earth, we need to work for food. We need to eat, as we shall be eaten one day. "We," *awi*, is ambiguous. Some people work more hours than others for a society to be reproduced materially and in language as a specific "we," and a whole humanity.[43] Nourishment is a condition for the possibility of all life, no matter what cosmic sphere it takes place on. Food is essential in Arawete ontology and thus in an Arawete theology of creation. To eat is to live. There is no creation prior to a being's becoming food through the activity of an eater's eating, and there is no created life for whom there was not already a living being to be eaten. Arawete ontology is relational. There is no being without another, no being eaten without an eater. If gods have always existed, so have many plants. The dialectic between eater and eaten is the fundamental form of both Arawete theology and Arawete society. The Arawete eat enemies. Women feed men. Eternal gods live opposite to finite humans. Gods eat humans. Humans serve food to the gods. Given that humans shall become gods, and gods were not differentiated from humans in paradise, the gods may be called "cannibals."

Cannibalism takes place among many kinds of beings across cosmic spheres, although not all beings are carnivores or cannibals. Some are vegetarian, although these lines become blurred when all beings have souls

---

42. John Bellamy Foster, inspired by Marx, calls it the "metabolic rift." Consumption has moved away from agriculture, and waste is not returned to nature's process of renewal. See Foster and Clark, *The Robbery of Nature*, 7–8, 42–46, 101–03, 215–17, etc.

43. *Bïde*, in Arawete, means both the pronoun "we" and Arawete humans, in contrast to enemies and gods.

that are the same in essence and different only in their carnal appearance.[44] To be a vegetarian is another way of being a cannibal. Every eating is a type of cannibalism. Patriarchy is a type of social cannibalism. Labor is a means of subsistence that, for the Arawete as for people under global capitalist production, has been organized by crashing relationships between classes of production. Under capitalism, the social disruption of classes has been extended from familiar relations of patriarchy and eventual wars between neighboring peoples to the whole of nature through the contamination of the environment and intensive resource extraction for profit. Patriarchy and speciesism are pivotal to global capitalism. As a predatory relationship of production, capitalism has relied on several other hierarchical social divides such as race, nation, and being able-bodied. If, on the one hand, the patriarchy to be noted among the Arawete may illuminate our global relations of gender-based production, Arawete food-centered theology serves as an anticapitalist critique insofar as intensive economic exploitation has disrupted the possibility of continuous reproduction of healthy food sources.

What difference does it make to the story of cosmic, planetary, human, and cultural evolution (as an adaptation to a place) to re-describe this as a product of the gods' ascension and humans' abandonment? What difference does it make to see humans in a temporary condition that can only regain the status of divine after a cannibalistic process of physical resurrection? What difference does it make to say that earth and heaven will once again be the same place and all the living shall share the same food from the forest? Because the Arawete aspire to become gods, they live a life re-described as a paradox. The pain and pleasure principles are intertwined in existence. Good and evil are inseparable until resurrection. Because this life used to be divine, it still holds the status of sacred. Insofar as the earth has been abandoned by the gods, life here has—borrowing a Christian image—become a cross. Finite life's condition of abandonment is not borne equally in society. Although all know suffering through different circumstances, women have been among the most exploited groups in many societies. Arawete creation theology explains, if not justifies, women's suffering. At the same time, Arawete theology highlights the new creations that can take place across cosmic spheres for as long as there is food. Food is condition of life. A neoliberal regime of fake food contradicts such a

---

44. According to Viveiros de Castro's description of Amerindian perspectivism, the reason why there are different perspectives is because we all think the same way from the point of view of different bodies. As a result of a diversity of bodies, the world is composed of a multiplicity of points of view. See Viveiros de Castro, *Cannibal Metaphysics*.

basic principle.[45] Not only the Arawete but many other traditions share the conception of creation from already-existing matter. The recognition of the dependence of all on the materiality of life could possibly also be turned against these traditions' own patriarchy once it is admitted that the inequality between the sexes was the beginning of all class division among humans and against nature.

■ Bibliography

Berg, Maxine. "What Difference Did Women's Work Make to the Industrial Revolution?" *History Workshop* 35 (1993), 22–44. https://doi.org/10.1093/hwj/35.1.22

De Beauvoir, Simone. *The Second Sex*, translated by Constance Borde and Sheila Malovany-Chevallier. New York: Random House, 2011 (1949).

"Enūma Eliš." In *Before the Muses: An Anthology of Akkadian Literature, Volume 1: Archaic, Classical, Mature*, edited and translated by Benjamin R. Foster, 353–401. Bethesda: CDL, 1996.

Federici, Silvia. *Wages Against Housework*. London: Falling Wall, 1975.

Fetterolf, Janell. "In Many Countries, at Least Four-in-Ten in the Labor Force Are Women." *Pew Research Center*, March 6, 2017, https://www.pewresearch.org/fact-tank/2017/03/07/in-many-countries-at-least-four-in-ten-in-the-labor-force-are-women/.

Foster, John Bellamy, and Brett Clark. *The Robbery of Nature: Capitalism and the Ecological Rift*. New York: Monthly Review, 2020.

Galeano, Eduardo. *Open Veins of Latin America: Five Centuries of The Pillage of a Continent*, translated by Cedric Belfrage. London: Latin American Bureau, 1997 (1971).

Hesiod. "Theogony." In *Theogony and Works and Days*, edited and translated by M. L. West, 1–33. Oxford: Oxford University, 1988.

———. "Works and Days." In *Theogony and Works and Days*, edited and translated by M. L. West, 35–61. Oxford: Oxford University, 1988.

Jácome, Camila, and Laura Furquim. "Teorias de Gênero e Feminismos na Arqueologia Brasileira: Do Dimorfismo Sexual à Primavera Queer." *Revista Arqueologia Pública* 13 (2019), 255–79. https://doi.org/10.20396/rap.v13i1.8654825

Jecupé, Kaka Werá. *Tupã Tenondé: A criação do universo, da terra e do homem segundo a tradição oral guarani*. São Paulo: Peirópolis, 2020.

King James Bible Online. Cambridge: Cambridge University, 2023. Last accessed February 25, 2023, http://www.kingjamesbibleonline.org.

Lee, Richard B. "What Hunters Do for a Living, or How to Make Out on Scarce Resources." In *Man the Hunter*, edited by Richard B. Lee and Irven De Vore, 30–48. Chicago: Aldine, 1968.

Marx, Karl, and Friedrich Engels. *The German Ideology: Including Theses on Feuerbach and Introduction to The Critique of Political Economy*. Amherst: Prometheus; New York: Random House, 1998 (1846).

Mies, Maria. *Patriarchy and Accumulation on a World Scale*. London: Zed, 1998 (1981).

Nimuendaju Unkel, Curt. *As Lendas da Criação e Destruição do Mundo Como Fundamento da Religião dos Apapocúva-Guarani*. São Paulo: HUCITEC/EDUSP, 1987.

---

45. Vandana Shiva puts it in numbers, stating of industrial agriculture that "it uses 75% of the land yet industrial agriculture based on fossil fuel intensive, chemical intensive monocultures produce only 30% of the food we eat. Meanwhile, small, biodiverse farms using 25% of the land provide 70% of the food. At this rate, if the share of industrial agriculture and industrial food in our diet is increased to 45%, we will have a dead planet. One with no life and no food." Shiva, "Fake Food, Fake Meat."

Oliveira, Marilza, "Mito Tupi-Guarani da Criação do Mundo." In *Danças Indígenas e Afrobrasileiras*, edited by Marilza Oliveira, 19-22. Salvador: Universidade Federal da Bahia, 2018.

Ribeiro, Berta. "Araweté: A India Vestida." *Revista de Antropologia* 26 (1983), 1–38. https://doi.org/10.11606/1678-9857.ra.1983.111039

Ruether, Rosemary Radford. *Gaia and God: An Ecofeminist Theology of Earth Healing*. San Francisco: Harper Collins, 1992.

Shiva, Vandana. "Fake Food, Fake Meat: Big Food's Desperate Attempt to Further the Industrialisation of Food." *Independent Science News for Food and Agriculture*. 2019. Last accessed February 25, 2023, http://www.independentsciencenews.org/health/fake-food-fake-meat-big-foods-desperate-attempt-to-further-industrialisation-food.

Vajskop, Amanda. "Finding Patterns in the Chaos: Woman as Chaos Agent in Creation Myths." *Denison Journal of Religion* 5 (2005), 60–73.

Viveiros de Castro, Eduardo. *Arawete: Os Deuses Canibais*. Rio de Janeiro: Jorge Zahar, 1986.

———. *Cannibal Metaphysics*. Minneapolis: Univocal, 2014 (2009).

Viveiros de Castro, Eduardo, Camila de Caux, and Guilherme Orlandini Heurich. *Araweté: Um Povo Tupi da Amazônia*. São Paulo: SESC, 2016 (1949).

The Struggle of Earth-Storytellers

Willie James Jennings[1]

■ Garden Life

I was formed in a garden, my mother and father's garden, which means that I was formed inside a story. I am the child of gardeners, which is to name a special grace—to be a descendant of people of the dirt, specifically Black people of the black dirt. My parents were sharecroppers in the South of the United States of America (USA) (the Jim Crow, hateful South) in the 1940s. In that South, which is not so far removed from the South of the USA at this moment, they lived their lives on two kinds of land, the land as torture and the land as relief, even promise. During what is known by United States (US) historians as the Great Migration, when Black Southerners left the American South in significant numbers heading north and west, my parents (who at the time had three children) came north, looking, in the words of Isabel Wilkerson, for warmer suns – we can add kinder suns as well.[2] They, like so many others who moved north, wanted to find some small patch of land on which to live and some factory job with which to build a living.

---

1. Willie James Jennings is an associate professor of systematic theology and Africana studies at Yale University. He is registered at the University of the Western Cape as a coresearcher for the project on "An Earthed Faith: Telling the Story Amid the 'Anthropocene'."

2. See Wilkerson, *The Warmth of Other Suns*. Also see Sernett, *Bound for the Promised Land*.

How to cite: Jennings, WJ 2024, "The Struggle of Earth-Storytellers", in EM Conradie & WJ Jennings (eds.), The Place of Story and the Story of Place, in An Earthed Faith: Telling the Story amid the "Anthropocene", vol. 3, AOSIS Books, Cape Town, pp. 155-169. https://doi.org/10.4102/aosis.2024.BK355.09

It is an important but overlooked fact that these Southern Black people (like European immigrants to America) wanted to live close to the land, not necessarily in cities but in places where they could farm a bit, or a lot, and where they could garden a bit, or a lot. This meant they wanted to live outside city centers, where folks with farming life in their blood could sense and know the ground. That dream, however, was killed by white people who did not want Black people near them. These hateful white folks living outside the hateful South created a new hell called sundown towns. These were towns where Black people who had moved there from the South were, through intimidation and violence, forced to flee into the cities. These were towns that created laws and geographic protocols that kept Black people out. Black people, chased out and blocked from living in these small towns with their larger plots of land, were forced to live tightly in environmentally dangerous parts of the bigger cities and towns.[3] This meant that Black folks moving north from the South often found themselves settling into places made into Black spaces.

My parents followed a Mississippi route to the northern part of the USA, through southern Illinois up to Chicago, and not finding Chicago to their liking, they continued north, but not all the way to Milwaukee, nor to Minneapolis, nor did they venture east to Detroit, but they settled in the western part of the Great Lakes state of Michigan (O beautiful Michigan!), in Grand Rapids, and there, after many years, I was born in time to see those gardeners in the eve of their lives tending their garden. My parents were very serious Christians, not the kind that so many people have come to despise in this horrific political moment of the Western world, but the kind of Christians that tended to people the way a good gardener tends to their plants—with kindness, care, generosity, and gracious attention. My parents were not perfect, but they were gardeners. I begin this essay with my parents because it shows the inextricability of their story from the story of life with the ground, life in the ground. Land and body are woven together in my parents' lives and in the lives of so many people. Yet the Christian doctrine of creation, as it has come to us and is yet often articulated by theologians, rarely touches the texture of bodies in land and their deep and abiding connection, as well as the struggle that has often come with that connection.

One of the fundamental flaws in modern Christian doctrines of creation is that they are obsessed with the question of origins.[4] I am not arguing in this essay that the idea of origins is unimportant, nor am I in any way aiming to continue the exhausted and misplaced fight between religion

---

3. See Loewen, *Sundown Towns*.

4. See Jennings, "Reframing the World."

(and Christian theology, specifically) and evolutionary theorization. One of my concerns is to release a Christian imagination from captivity to the interface with evolutionary thought, as though that is the first and most important conceptual matter we should be concerned about in reflecting on creation. This essay presses against one of the central questions of this volume:

> What difference does it make to the story of cosmic, planetary, human, and cultural evolution to re-describe this as the creative work of God's love? Inversely, what difference does it make to the story of God's love to describe it in evolutionary and geographic terms?

These questions constrict because they press us to see the world first in evolutionary terms; that is, they invite us to attend to the world inside a temporal logic rather than a spatial logic.

The question I explore in this essay is what does it mean to attend to the world inside a spatial logic, or more precisely, inside the logics of place and habitation? The problem we face as Christian theologians and intellectuals is that our conceptualities in relation to the world and a doctrine of creation are controlled by temporal logics which have made us easily susceptible to a certain kind of intellectual conceit in our thinking.[5] We too quickly move to ways of thinking that displace us, positioning our imaginations beyond the specifics of place. Evolutionary theorization lends itself to aiding our displacement. Again, I need to state that I am not denying evolutionary vision, nor am I ignoring the truth that evolutionary theorization does in fact require attending to biological details and has yielded stunning insights into the nature of the world.[6]

I want to consider what it means to take place seriously in a Christian doctrine of creation. To take place seriously requires we take seriously the earth-storytellers, earth-storytellers like my parents, and therefore to take seriously the struggle of earth-storytellers. This means that a doctrine of creation must begin with the crucial question of "who" is the storyteller before we engage "how" the story is told or even "what" story is told. Organizing these questions in this manner turns our focus to the relation of the body to land and the struggles that exist precisely at this crucial nexus. Centrally, that brings us into the struggles of women, especially women of color, in the world who consistently inhabit these sites of struggle. By situating the earth-story struggle with women of color, I am by no means following a colonial logic of equating women with the earth as one

---

5. See Sideris, *Consecrating Science*; also Slade, *The Fullness of Time*.

6. See Gould, *The Structure of Evolutionary Theory*. Also see Cavanaugh and *Evolution and the Fall*; Deane-Drummond, *Christ and Evolution*; Ruse, *The Gaia Hypothesis*.

continuous body in need of claiming, conquering, and cultivating. Nor am I necessarily engaging the so-called Gaia hypothesis.[7]

My focus is on the struggle over earth-storytelling that involves what Silvia Federici and Vandana Shiva, in different ways, articulate as the fight over the seed, both in terms of the control over the production of food and the reproduction of life, that so many women have been engaged in and continue to be engaged in all over the world.[8] Yet the struggle itself carries hermeneutical import in that it illumines the story of creation in the creation of story, that is, in the ways women continue to announce what is necessary for thriving life through their stories. The storyteller, in this way, emerges as one who speaks of creation and one through whom creation itself speaks. So, my way of proceeding in this essay will be to briefly consider the problem of displacement and then turn to the body–land connection as the way to think of creation at the sites of struggle.

■ Displacement

The Native American religious scholar Vine Deloria Jr. is credited with many things, including being one of the first religious intellectuals in North America to articulate the problems in Christian theology that are exposed by the history and reality of Indigenous life. Through several groundbreaking books, including *Custer Died for Your Sins* (1969) and the epic text *God Is Red: A Native View of Religion* (1973), Deloria showed the conceptual horrors embedded in the physical horrors inflicted on Indigenous people by European colonialists.[9] One of the least considered damages was imposing a form of temporality that denied spatiality. Two crucial colonial agents facilitated this imposition, the merchant and the missionary. By merchant, we mean more than the buyer and seller of goods and services. Merchant in this regard means all those involved with the seizure of indigenous land, with its subsequent fragmentation, commodification, buying, and selling. The merchant denied the way Indigenous people named, identified with, and gained identity from places. In short, the merchant denied place-centered identities.[10]

---

7. See Merchant, *Reinventing Eden*; also her *Science and Nature*.

8. See Federici, *Caliban and the Witch*; also her *Revolution at Point Zero*; Shiva, *Staying Alive*; also her "The Impoverishment of the Environment."

9. Deloria, *Custer Died for Your Sins*; *God Is Red*; also *Spirit and Reason: The Vine Deloria Jr. Reader*. Also see Churchill, "Contours of Enlightenment: Reflections on Science, Theology, Law, and the Alternative Vision of Vine Deloria Jr."

10. See Jennings, *The Christian Imagination*.

Moreover, merchants turned place into space; that is, they turned places into simply land that now carried a new future—it could and would be owned and developed. Ownership and development are what situated land inside a new temporality that eradicated any history that might dictate the forms and character of engagement with that land and the plants and animals that inhabited that land. In this regard, land includes bodies of water and waterways, fishing grounds, and excavation sites. The denial of land's history in this regard is the denial of its power to emplace people, that is, to center identity and consciousness itself in a logic of habitation. The missionary, in relation to the merchant, was one who desacralized place for the purpose of creating a sacred space. Missionaries throughout the colonial theaters challenged the ideas, frameworks, conceptualities, and practices of Indigenous peoples who saw the land and animals as animate and communicative, as persons, even though they were not human.[11] The land for the missionaries was *terra nullius*, both in terms of ownership and divine presence. The point of rehearsing a bit of this history is to capture the dilemma that Vine Deloria Jr. so powerfully noted—that Christianity presented to native peoples the idea that time was the register to think of creation rather than space, and certainly not place. Created reality in this schema is intelligible and meaningful through transformation rather than habitation.

The idea of transformation captures several concerns of merchant colonialists—development, extraction, production, and commodification—and of missionary colonialists—conversion, spiritual growth, and cultural assimilation. What tied these concerns together were laboring bodies. Here, we have the foundational logic of all colonial terrorism with its theft, enslavement, and exploitation. But we also have the foundations for a view of land that resists the logics of place. By the logics of place, I mean not simply life in a place but life with the ground and animals, life in reciprocity. Reciprocity in this regard is not a romanticism; it is the varied protocols of living in places where the negotiations of predation, planting and harvesting, hunting and fishing, and growing and protecting food and life reveal a deep sense of the web of living and our enmeshments with all creatures.[12] We have inherited these colonialist trajectories, which have hollowed out the sense of habitation for so many people in the world. Equally important for our considerations, they have hollowed out the

---

11. See Tinker, *Missionary Conquest*; Viveiros de Castro, *The Inconstancy of the Indian Soul*; Harvey, *Animism*.

12. See Abram, *The Spell of the Sensuous*; Kohn, *How Forests Think*; Wirzba, *The Sacred Life*.

sense of the spatial and the reality of place in the way we think about a doctrine of creation.

The anthropologist Tim Ingold speaks of this hollowing phenomenon in terms of the problem of inversion. Inversion is a condition of perception and life whereby people envision their lives as enclosed realities moving along the surface of the ground, moving from space to space, like a car on a highway. Inversion means we imagine that we live on the surface of the ground and not through and with the ground. We live moving from spot to spot as if on a map. For Ingold, we are not enclosed realities (like circles rolling on a straight plane); we are lines that move through places, leaving traces, and in so doing, we cross other lines (other lives). What results from that constant crossing is a meshwork.[13] A meshwork is not a network, which suggests separate entities interacting. Meshwork means that who we are and how we are formed is the constant movement through places, going out and returning, and in so doing, engaging people. Meshwork also means that there is a reality of connectivity to place that can be denied or acknowledged, made life-giving or life-destroying, conducive to destructive forms of identity or supportive of interwoven forms of identity that expand into the ground, air, and landscape.

Our sense of dwelling in terms of meshwork, then, is critical to how we understand life in the world and for centering the logic of a doctrine of creation. It is also crucial for understanding the long legacy of struggle for earth-storytellers. Earth-storytellers in this regard are those people who have fought against this reality of displacement that seeks to make them floating bodies living on the surface of the land, whose labor is extracted and exploited, and who are being pressed to live alienated from the land they inhabit, even if they do or do not own it. This, in fact, is the plight and fight for so many women of color in the world. The Ghanaian womanist theologian Mercy Amba Oduyoye narrates this struggle in terms of African women being pressed into patriarchal logics around land ownership and cultivation, where centuries-long practices of women in guiding and controlling the means of food production, food creation, and seed cultivation, as well as processes of exchange and market participation, have been undermined and, in many cases, resisted and destroyed. She notes that women's lives are inextricably bound to the exploitation of the land.

> When a poor country has to export more to already rich countries, it takes land from the poor, especially women, to grow what the North needs, not what mothers in the South need to feed children. When governments cut spending,

---

13. See Ingold, *Being Alive*; also his *Lines* and *The Life of Lines*.

schooling and healthcare fall on families and all work triple-time just to be able to feed the children—so mothers eat last.[14]

Vandana Shiva, philosopher and physicist, notes for us the ways women are being criminalized for opposing the takeover of their agricultural practices by corporations that steal both their knowledge of seeds and the seeds themselves and then claim proprietorial rights over those same seeds.[15] Modern industrial societies forming in the energy of patriarchy and the impulses of colonialism and capitalism devalued the processes of regeneration, which Shiva reminds us "is the cause of both the ecological crisis and the crisis of sustainability."[16] As she states:

> The continuity between regeneration in human and nonhuman nature that was the basis of all ancient worldviews was broken by patriarchy. People were separated from nature, and the creativity involved in processes of regeneration was denied. Creativity became the monopoly of men, who were considered to be engage in production; women were engage in mere reproduction or recreation, which, rather than being treat as renewable production, was looked upon as nonproductive.[17]

There is much to unpack in this quote, but the most relevant to our concerns is the insight regarding the importance of the processes of reproduction and regeneration. It is precisely the struggles over acknowledging, sustaining, and protecting these processes embodied in the lives of people, especially women, that bring us most centrally to the connection of the body to land. These struggles point to the place from which our storytelling must begin. We tell the story of creation from the position of those seeking to sustain a connection to the earth in the life work of regeneration. More precisely, we listen to their telling of the earth-story in their story, and in so doing, we are also hearing the earth (in specific places) speaking with and through them. Allow me to return briefly to my own story.

■ The Body-Land Connection

I grew up in the seventies in a Black community circumscribed by white control, which is to say I grew up in the kind of Black community that characterizes much of America. The house I grew up in was in a neighborhood that had recently been made Black by White flight, the ingenious practices of redlining (tightly controlling where Black people could seek to buy homes and live), and the courageous efforts of Black folks to make

---

14. Oduyoye, *Breads and Strands*, 62; also her *Daughters of Anowa*.

15. See Shiva, "Monocultures of the Mind," in *The Vandana Shiva Reader*, 71–112. See also Shiva and Mies, *Ecofeminism*.

16. Shiva, "The Seed and the Earth," in *The Vandana Shiva Reader*, 159.

17. Shiva, "The Seed and the Earth," in *The Vandana Shiva Reader*, 159.

enormous sacrifices, save their money, accept unfairly high interest rates, and move into a house made theirs by faith and hope. My parent's home at 717 Franklin Street was a house on a row of homes owned or rented by Black folks, and in the back of each home was a garden (some larger, some smaller). On a summer day, standing in the backyard, your senses were treated to a culinary festival—tomatoes, cucumbers, apples, peaches, blackberries, cherries, melons, beans, and everything the rich black Michigan dirt could womb to life.

The gardens were places located geographically and spiritually between heaven and church. In that holy position, they served two abiding and beautiful functions. The first was that they made the informal economy work. This was ancient stuff—knowledge that goes back to the beginning of community, where practices of exchange were in the service of community and the building and sustaining of community, not the other way around, where communities' farming practices are made slaves to the greed of global exchange. I grew up in a large family—nine of us living at 717 Franklin—and my father had a good job, but even that good job could not make ends meet without the economy of the garden. And here is how that economy worked.

"Willie, come here, boy. Take this basket of tomatoes next door to Mrs. P., and bring those eggs she has for me from her brother's garden," she would say, because he has chickens.

"Yes, ma'am."

"And when you get back, I want you to take these cherries over to Mother H., she has fresh fish for us."

I walked and biked all over the community, delivering and picking up food. Then there came that moment when you felt the first thin chills in the approaching winter wind, and you spotted the first turning of the leaves. At that moment, you knew it would soon begin, the canning, the jars, the preserving—things got serious because, like squirrels and chipmunks and birds, we were preparing for winter. The seriousness of that fall preserving would be seen in the dead of winter because, in the middle of a cold Michigan winter, you could have fresh peaches, apples, tomatoes, or blueberries in pie or made into a jam.

We survived because a garden was joined to a job. But there was another joining that showed the second abiding and beautiful function of the garden. When the Michigan winter gave way to the warmth of late spring and summer and early fall, brief though it was, you would find my parents and so many other Black folks who knew the South, and what it was like to feel the good earth beneath their feet, doing a very wise thing. They would be sitting out back (as we say) in old cast iron chairs, the ones with seats that almost touched the ground, and there they would enjoy being outside and left free to be. In those backyard moments of heavenly relief, there would be talking and laughing and shouting and singing across the fences to their neighbors—the back-and-forth signaling the freedom to speak the truth about life and love and struggle to one another. More importantly, those moments also revealed a symbiosis and a holy surrogacy through which trees and plants, bushes and animals, earth and sky, and Michigan dirt spoke through these Black elders of the earth. They inhabited an indigenous logic and sensibility that the world tells its own story through us, that creatures speak with and through creatures.

Every moment was not heaven, but there were moments where heaven's foretaste could be sensed. Yet my parents and the other elders of the dirt were in a place that we, their Michigan children, had real difficulty entering, because we were on a different journey. We lacked the pedagogy of the dirt that they had won through ancient remembering mixed with hard life in the South. We were Northern children in a highly segregated, profoundly sequestered reality of free life; that is, you could go anywhere you wanted to go, but there was really no place to go and few, if any, places that wanted you to come. The seventies were the days of *forced* school busing and *macro*-aggressions against Black kids in schools, especially middle and high school, and the beginnings of aggressive policing, growing surveillance, and cities that formed social, cultural, political, and intellectual life as though Black people did not exist.

We felt it—in our bodies, because we felt it in our places, in the land itself, menacing our souls and the ground. What do you feel when you go to school and every day the vast majority of your teachers see you and the place you live through a demeaning and derogatory veil through which they invite you to see life and your body? What do you feel when you look at the newspaper or watch local television and 99% of everything that is presented as true, as good, as beautiful, as noble, as talented, and as announcing the future is white and inhabiting white spaces? What might you feel when the only life options being imagined for you by many of the adults around you and almost all the white adults you know from school are either the military or a factory job, and for a very few there was junior college? What do you feel when the place you inhabit is sequestered, strangled of natural beauty—few if any parks, swimming pools, green spaces—and the ones you aim to go to are in heavily guarded and surveilled places? These questions point to the struggle to be and become earth-storytellers in the face of the continued legacies of colonialism remaking the world into a commodity.

If the story of creation (as Christians might tell it) must begin with the struggles of the earth-storytellers, then what stands over against this positionality are those ways of telling the story of creation that begins with the earth as an object. The earth as object presents a form of displacement through which the world may be analyzed in ways that extract knowledge of its inner workings. Displaced analysis yields displaced storytellers who imagine themselves suspended above space and time with a god-like view of everything.[18] Lisa Sideris, in her *Consecrating Science: Wonder, Knowledge, and the Natural World,* examines the effect of this displacement of story and storytellers in the ways scientific exploration has been supplanted by some thinkers with scientism. Scientism, in this regard, is a philosophical framework that claims (Western) science as the only true intellectual plateau from which to see the truth of the world and scientists (or commentators on scientific pronouncements) as the only arbiters of truth.[19] Sideris seeks to challenge scientism through reasserting the idea of wonder. She understands wonder to be the sense of awe that provokes humility and openness in the presence of mystery. Sideris wants to re-establish wonder as a basis on which scientific exploration moves forward, and wonder as the framework within which knowledge is sought, produced, corrected, and built upon, rather than a vision of wonder being understood as the result of scientific discovery.

Sideris notes the way wonder fell into disregard inside the disenchantment of the world. That disenchantment meant that the world was no longer seen as animate and communicative, nor permeated with spirits. Wonder was associated with ignorance and primitivity, that is, wonder was no longer seen as having anything to do with knowledge and truth-seeking.[20] There are dense colonial histories behind the emergence of a framework of disenchantment through which to understand the world. For our purposes, that emergence deeply affects the ways we understand and tell the earth-story. Sideris also notes how wonder and curiosity were severed and curiosity re-emerged as a virtue of the scientific method. This was a result, in large measure, of the way curiosity was positioned as a technology of acquisition.

---

18. See Sousa Santos, *Epistemologies of the South*; also his *The End of the Cognitive Empire*. Sousa Santos calls what I term displaced analysis a form of abyssal thinking, that is, thinking that enacts a system of visible and invisible distinctions that separate reality into what exists and what does not exist; in this case, the thinking of the Global South does not exist and the thinking of the Global North not only exists but constitutes existence and in this way prepares us to engage in intellectual extractive practices in the world; see *Epistemologies of the South*, 118–63.

19. Alister McGrath defines scientism as "a contracted version of 'scientific imperialism' […] understood as 'a totalizing attitude that regards science as the ultimate standard and arbiter of all interesting questions; or alternatively that seeks to expand the very definition and scope of science to encompass all aspects of human knowledge and understanding'." See McGrath, *Re-Imagining Nature*, 156. For this definition, McGrath is drawing on the work of Massimo Pigliucci, "New Atheism and the Scientific Turn in the Atheism Movement," as well as Roger Scruton, "Scientism in the Arts and Humanities."

20. Sideris, *Consecrating Science*, 14–28.

Curiosity, in this regard, does not require any narrative framework within which to understand its logic. It is its own logic. But what is also the case is that curiosity was an important engine in creating commodities. We should think of this reframing of curiosity through one crucial question: "Is there a market for this?" That question, asked by centuries of merchants, drew curiosity toward its captivity to capital.

Sideris also notes the formation of a cosmic story of the universe (executed primarily by commentators on science) within which we are being offered a basis in an authentic reading of reality. The real cosmic story of the universe is presented by such commentators as the proper basis on which to engage in mythopoetic thinking about the world, which is precisely what religious thinking about the world is in this schema. In this regard, religion and science are in competition regarding truthful accounts of existence. Science has won and offers religion an important role in articulating the truths *of its victory*. The cosmic story becomes the basis for imagining a shared project of living rooted in a shared project of knowing the truth of existence. As Sideris outlines it, wonder returns in this scientism schema as that which the scientists show us how to see. She is tracking the hubris that hides in the false humility of scientific pronouncements about the nature of all existence. Wonder is constituted in the results of scientific discovery, and therefore, the sacrality of the world comes into view only after scientific investigation has taught us how to see it for what it is.

Sideris sees it as a dubious endeavor to think we can form an environmental ethic that has positioned wonder and, by implication, sacrality as the result of scientific investigation and pronouncement. In fact, trying to think of the world in the cosmic register opens us up to two problems: firstly, we situate human beings at the very center of the story, with a god-like view of everything, and secondly, we ultimately see the world in fragmentation as a collection of things like vibrant particles. In other words, we will tend toward not taking the lives of the more-than-human world seriously as living others that confront us with the question: how will we live, and how will we live together? More importantly for our concerns, the positionality presented to us in this form of earth-storytelling follows the colonialist trajectories we noted earlier and loses sight of the earth-storytellers and a practice of storytelling that allows the earth to speak through us.

My goal in this essay is not to dismiss scientific pronouncement or limit forms of scientific investigation. Rather, I wish to frame those activities inside the work of storytelling that cannot be taken away from the peoples of the earth. I want to frame the work of earth-storytelling by naming the ignored subject and noting the ways in which that ignored subject has been and continues to be concealed in and through the kinds of storytelling that Christian doctrines of creation tend to promote. Yet the subject here is twofold: on the one hand, women of color, and on the other hand, the earth itself in the specifics of places. I am positing neither a metaphysical nor a gender-exclusive symbiosis in earth-storytelling, rendering women as women to be closer to nature. Rather, I am seeking to supplant temporal logics for storytelling with spatial logics. In alignment with scholars like womanist theologian Karen Baker-Fletcher and ethicist Melanie Harris, I am asking what it would mean to begin with our placement in the specific struggles of existence and expression for land and body.[21] As Karen Baker-Fletcher notes, in her text, *Sisters of Dust, Sister of Spirit*:

> If we forget we are people of the land, who belong to God, we lose connection to our own spirits and to God as the life-force of Creation. Care of the land is interrelated with care for our bodies, minds and spirits. We must awake from the seductive sleep of spiritual torpor and remember who we are.[22]

This repositioning of earth-storytelling firstly frames earth-storytelling inside the struggle of life and the struggle to maintain the web of life, and secondly, it yields cognitive and affective space to land and animal, water and sky, plant and wind to speak themselves through our speaking. In effect, this repositioning helps us enact a commitment to an animate and communicative world. Such a commitment takes seriously the symbiotic relation of humans and nonhuman persons, resisting the habit of mind that seeks to isolate planetary struggle from human struggle, and the specific plight of species extinction and the stripping away of habitation from Indigenous communities and the loss of human life. That habit of mind is rooted in a colonialist optic that separates the being of the land from the being of people. We must therefore think through the problem of centrism from the ways it formed through a colonialist optic.

What does it mean to be centric? As I have noted elsewhere, the centrism of modern colonialism begins with its supersessionism.[23] Proto-European colonial Christians imagined themselves as replacing and supplanting the Jewish people at the center of the divine will in all aspects of existence—how one works, how one lives, how one learns, and how one sees both time and space. This is a centrism that is both ontological and pedagogical, both of being and how to form human beings. It is a centrism that, as it grew in the new worlds woven inside their colonial theaters, spread beyond Christian conceptual frameworks but sustained this Christian framing.

---

21. Baker-Fletcher, *Sisters of Dust, Sisters of Spirit*; Harris, *Ecowomanism*. Baker-Fletcher and Harris, in different ways, continue to note the relation of Black women's lives and the life of the planet, as well as the interconnectedness of Black thriving to the thriving of the earth.

22. Baker-Fletcher, *Sisters of Dust, Sisters of Spirit*, 58.

23. See Jennings, *The Christian Imagination*, Chapter 1.

This centrism carries many abiding characteristics that are with us now, but there are two that concern us for our purposes: the refusal to center nonwhite peoples in their ways of knowing, being, and living with the earth, and the positioning of the European between people and the land as the universal arbiter, evaluator, and judge of what is best for both people and land. The epistemological frameworks that have emerged from this centrism not only determine what counts as knowledge but also what constitutes authentic insight into the world. This is centrism that forms Eurocentrism, human-centrism, androcentrism, Western-centrism, and white supremacy. Yet it also creates a form of critique—global critique—that flows out from this centrism, which conceals itself as a form of centrism, even as it offers analysis and critique of worldwide problems. We see this in nation-states and corporations that seek to tell the peoples of the world how to organize or reorganize their ecologies for maximum productivity and profit. We can also see this in policies and strategic plans that isolate people from land, removing people from places for the sake of "saving" animals, flora, and fauna or for the sake of seizing natural resources. Centrism in this regard, then, is a conceptuality rooted historically in displacing Indigenous people's voices and vision of life in their land and abstracting land, water, animals— their life and their well-being—from the opinions, ideas, plans, and practices of the very people who inhabit that land and live among its animals.

The problem of human-centrism therefore requires proper historical nuance to understand it as a problem. It is not the centering of humans to the detriment of plants, animals, and the ecosphere. It is the centering of *some* humans who stand squarely in the legacies of advantage, both ontological and pedagogical, that positions their voices, their vision, their policies, and their practices as determinative for the whole planet. Human-centrism is fundamentally white masculinist centrism and the inertia of global harm activated by that reality of centrism that captured and continues to draw many peoples into ecologically destructive practices for the sake of participation in a global market. Addressing this problem begins with a new practice of storytelling that moves inside the spatial logic of place and emplacement, listening to the voices of women of color who count time through dirt-bound life. Might it be better to see time through the earth? By this question, I do not mean radiocarbon dating (or carbon-14 dating) of the earth. I mean seeing the time we are in through the realities of habitation, allowing place to situate us in the time of God.

This would require taking hold of life in place as a form of concave living rather than convex existence. Karl Barth, in his famous commentary on the book of Romans, wrote of the church as always being in danger of imagining itself as a form of convex existence, that is, an existence that offers to the world a truth about God and about life that it contains within itself, rather than being a form of living that is concave, always an opening, a crater, waiting for divine light and divine life to fill us and move through us.[24] I would like to expand Barth's insight here to speak of life as part of creation. We are creatures called to a form of concave living that allows for a filling that is bound to another filling, where the creation itself sings through us with the voice of God speaking to and through us. This is the difference the story of God makes for how we understand creation in all its cosmic, scientific, and evolutionary dimensions. The other creatures—wind and rain, dirt and plant, seasons and waters, sky and moon, sun and stars—sing like a choir through us as God sings the lead. Thus, our storytelling of God requires earth-storytelling and, most importantly, listening.

■ Bibliography

Abram, David. *The Spell of the Sensuous: Perception and Language in a More-Than-Human-World*. New York: Vintage, 2017.

Baker-Fletcher, Karen. *Sisters of Dust, Sisters of Spirit: Womanist Wordings on God and Creation*. Minneapolis: Fortress, 1998.

Barth, Karl. *The Epistle to the Romans*. London: Oxford University Press, 1968.

Cavanaugh, William T., and James K. A. Smith, eds. *Evolution and the Fall*. Grand Rapids: Eerdmans, 2017.

Deane-Drummond, Celia. *Christ and Evolution: Wonder and Wisdom*. Minneapolis: Fortress, 2009.

Deloria Jr., Vine. *Custer Died for Your Sins: An Indian Manifesto*. Norman: University of Oklahoma Press, 1969 [Reprinted, 1988].

———. *God Is Red: A Native View of Religion*. Golden: Fulcrum, 1973 [Reprinted 2003].

———. *Spirit and Reason: The Vine Deloria Jr. Reader*, edited by Barbara Deloria, Kristen Foehner, and Sam Scinta. Golden: Fulcrum, 1999.

Federici, Silvia. *Caliban and the Witch: Women, the Body and Primitive Accumulation*. Brooklyn: Autonomedia, 2004.

———. *Revolution at Point Zero: Housework, Reproduction, and Feminist Struggle*. Brooklyn: Autonomedia, 2020.

Gould, Stephen Jay. *The Structure of Evolutionary Theory*. Cambridge: Harvard University Press, 2002.

Harris, Melanie L. *Ecowomanism: African American Women and Earth-Honoring Faiths*. Maryknoll: Orbis, 2017.

Harvey, Graham. *Animism: Respecting the Living World*. New York: Columbia University Press, 2006.

Ingold, Timothy. *Being Alive: Essays on Movement, Knowledge and Description*. New York: Routledge, 2011.

———. *The Life of Lines*. New York: Routledge, 2015.

———. *Lines: A Brief History*. New York: Routledge, 2016.

Jennings, Willie James. *The Christian Imagination: Theology and the Origins of Race*. New Haven: Yale University Press, 2010.

———. "Reframing the World: Toward an Actual Christian Doctrine of Creation." *International Journal of Systematic Theology* 20:4 (2019), 388–407.

---

24. Barth, *The Epistle to the Romans*, 149–87.

Kohn, Eduardo. *How Forests Think: Toward an Anthropology Beyond the Human*. Berkeley: University of California Press, 2013.

Loewen, James W. *Sundown Towns: A Hidden Dimension of American Racism*. New York: Touchstone, 2005.

McGrath, Alister E. *Re-Imagining Nature: The Promise of a Christian Natural Theology*. Malden: Wiley Blackwell, 2017.

Merchant, Carolyn. *Reinventing Eden: The Fate of Nature in Western Culture*. New York: Routledge, 2003.

———. *Science and Nature: Past, Present, and Future*. New York: Routledge, 2018.

Oduyoye, Mercy Amba. *Beads and Strands: Reflections of an African Woman on Christianity in Africa*. Maryknoll: Orbis, 2004.

———. *Daughters of Anowa: African Women and Patriarchy*. Maryknoll: Orbis, 2000.

Pigliucci, Massimo. "New Atheism and the Scientific Turn in the Atheism Movement." *Midwest Studies in Philosophy* 37:1 (2013), 142–53.

Ruse, Michael. *The Gaia Hypothesis: Science on a Pagan Planet*. Chicago: Chicago University Press, 2013.

Scruton, Roger. "Scientism in the Arts and Humanities." *New Atlantis* 40 (Fall, 2013), 33–46.

Sernett, Milton C. *Bound for the Promised Land: African American Religion and the Great Migration*. Durham: Duke University Press, 1997.

Shiva, Vandana, and Maria Mies, eds. *Ecofeminism*. London: Zed, 2014.

———. "Monocultures of the Mind." In *The Vandana Shiva Reader*, 71–112. Lexington: University of Kentucky Press, 2014a.

———. "The Seed and the Earth." In *The Vandana Shiva Reader*, 159. Lexington: University of Kentucky Press, 2014b.

———. *Staying Alive: Women, Ecology, and Development*. Berkeley: North Atlantic, 2016.

Sideris, Lisa. *Consecrating Science: Wonder, Knowledge, and the Natural World*. Berkeley: University of California Press, 2017.

Slade, Kara. *The Fullness of Time: Jesus Christ, Science, and Modernity*. Eugene: Cascade, 2021.

Sousa Santos, Boaventura. *The End of the Cognitive Empire: The Coming Age of Epistemologies of the South*. Durham: Duke University Press, 2018.

———. *Epistemologies of the South: Justice Against Epistemicide*. New York: Routledge, 2014.

Tinker, George E. *Missionary Conquest: The Gospel and Native American Cultural Genocide*. Minneapolis: Fortress, 1993.

Viveiros de Castro, Eduardo. *The Inconstancy of the Indian Soul: The Encounter of Catholics and Cannibals in 16th Century Brazil*. Chicago: Prickly Paradigm, 2011.

Wilkerson, Isabel. *The Warmth of Other Suns: The Epic Story of America's Great Migration*. New York: Vintage, 2010.

Wirzba, Norman. *The Sacred Life: Humanity's Place in a Wounded World*. Cambridge: Cambridge University Press, 2021.

Continuing the Conversation on Creation in Christian Ecotheology

1. **EMC and WJJ: The series on *An Earthed Faith* draws on various strands of narrative theology from around the world. The assumption is that Christian faith has a narrative shape and structure. It tells a story—interpreted differently by different people—of who God is and what God has done, is doing, and may do within our world. Do you have any comments in this regard, given the various contributions to this volume?**

SC: I find this volume intriguing particularly because it presents a platform for contributors from their varied leanings on the Christian faith to reflect on the creative work of God. The creation motif is tackled not only from the Genesis story of creation, but also from indigenous traditions. My essay juxtaposes the Genesis creation stories with the Indigenous Shona myths of creation. I also foreground the pertinent need for a gendered reading of the creation myths. In my view, most African Christian communities have been heavily influenced by the Indigenous creation myths; hence, our analysis of the Christian creation stories should also take cognizance of how the practicing Christian communities retrace their origins through their Indigenous narratives.

How to cite: Conradie, EM, Jennings, WJ, Guðmundsdóttir, A, Jacobsen, E, Andrianos, L, Harris, ML, Aldred, R & Chirongoma, S 2024, 'Continuing the Conversation on Creation in Christian Ecotheology', in EM Conradie & WJ Jennings (eds.), *The Place of Story and the Story of Place,* in An Earthed Faith: Telling the Story amid the "Anthropocene", vol. 3, AOSIS Books, Cape Town, pp. 171–183. https://doi.org/10.4102/aosis.2024.BK355.10

EJ: Why do religious beliefs have a narrative structure? Could it be because that is how we think about ourselves as we happen in time? Because to think of oneself as different from the divine is our way of thinking of our humanity? Narration does not exclude reason (Habermas). All language is the thought of alterity (Kant, Hegel, Levi-Strauss, Levinas, Beauvoir, Viveiros de Castro). To think is to think oppositively. To be is to do actions of contradiction, as no action is the same as another. For the Arawete, eater and eaten are the essential conflictual relationships in life, which does not take place without death. Thinking is the mental materialization of class struggle. The human condition is a paradox oscillating between pleasure and pain. Christians often wonder: Why did God create both good and evil if God means well? Is creation a process still taking place? The Arawete do not ask that question because the gods created neither good nor evil. Paradise and forests have simply existed from the beginning of any time. The fall from primeval peace, absence of labor, and plentifulness of food and music was an accident, not a creation. The Arawete rather ask, like Christ, why the resurrecting gods have abandoned them to die if the Arawete are meant to live forever. Nobody understands suffering, for it has no reason to be. Or does it? Are the robbed lifetimes of colonized, racialized, and genderized workers what history demands for the creation of justice?

LA: Christian faith has a narrative structure as it relates to one's experiences of living with God as the mighty controller of everything. Because every individual has different characteristics and responses to God's plan, faith is expressed differently in a narrative way. This depends on a personal understanding of every circumstance.

AG: The beauty of this volume is that it is a collection of very different stories from across the globe. To be able to tell our own story and to place it next to other stories makes it so obvious that we all have unique stories to tell, while we, at the same time, all belong to one big story. Our many stories are only pieces in the big puzzle, where no one piece is more important than the others. We need all of them in order to see what the picture looks like, just as we need a whole collection of stories in order to complete the one story.

MLH: Stories can be shared as an act of powerful resistance. In this way, the weaving of story and place in this volume is, in fact, a declaration that the time in the "Anthropocene" has come for the undoing of colonial frames in ecotheology. By inviting deep reflection on space, place and creation stories, this volume remaps the theoretical frameworks and scrambles the theological categories that have been used to keep normative claims of dominance in place. For those who have ears, let them hear.

RA: Most of my writing ends up trying to hold together Indigenous identity and Christian faith, and this happens in the form of a story. Creation is the context where all this happens. I wrote about treaty because Indigenous treaty was about making a covenant between different peoples and the Creator—so that we would have peace. If we could understand that we are related to Earth and to all things, we could live better stories. Creation theology is thus central because creation is where we experience the Creator, and all of our experiences are human.

2. **EMC and WJJ: For this volume, the focus is on God's work of creation. The question that we addressed was this: "What difference does it make to the story of cosmic, planetary, human, and cultural evolution to re-describe this as the creative work of God's love?" Inversely, what difference does it make to the story of God's love to describe it in evolutionary and geographic terms? What were the crucial insights that emerged for you as you encountered the essays of the other authors and worked on your own essay?**

RA: Creation is the context for all of our encounters with God (the Creator). This means, for Indigenous people, that all of creation is sacred, for the Creator is doing something powerful in and through creation. I think our problem as human beings is that we do not always understand ourselves as part of creation. There is no life apart from creation. I think it was Barth who pointed out that all we know about God is what we understand through revelation. Indigenous people would agree and see that revelation around them. Creation then becomes the context where we join with the Creator to work to see Creator and creation in perfect harmony. As creation is the context of all stories, we needed a book with many perspectives to add to the story.

LA: The stories from other authors—and my own story—are expressions of the abundant love of God in unlimited forms. This volume is a marvelous chance to admire the richness of God's creation geographically and spiritually. It is also an opportunity to depict the various ways of explaining Christians' faith in different contexts and places. God's creation is a continuous work involving all Christians as coworkers for building the kingdom of justice and peace on Earth as it is in Heaven.

SC: Reading through the essays written by the other contributors and thinking through the Shona creation myths, which are the focus of my essay, it became apparent to me that although both the Genesis creation myth and the Shona myths of origin bring to the fore that the universe was created as not only "good," but "very good," the existence of evil is an inevitable reality. The contributions in this volume also restate that the existence of evil does not annul God's goodness and lovingkindness. This volume has also further enhanced my conviction that our interaction with the created universe has a huge bearing on how Mother Earth relates to us. I have also been further enlightened to see that the occurrence of some natural disasters is not always because of our human misdeeds but an act of nature. Hence, as part of the created universe, we are continually reminded that God has a grand purpose for the universe, and some of the mysteries of nature will remain hidden from our purview.

EJ: I proposed a noncreation theology because Arawete true gods do not create out of nothing. Nonetheless, Arawete true gods are responsible for what Christians might consider God's greatest act of creation: the resurrection of all the dead. Why do the Christian God and the Arawete gods resurrect the dead if not out of love? To have a resurrecting God means, not only for Christians, to be loved even when abandoned by everyone else.

MLH: One crucial insight that emerged for me in writing my own essay and reviewing others is how important the arts are as vehicles for expression, especially when listening to the voices of women engaging in Earth justice and hearing the voice of Earth. In my essay, for example, it should not be lost to us that Fannie Lou Hamer was a singer connected to the earth and that James Weldon Johnson's art includes bountiful images of Earth.

AG: What became obvious for me, as I was starting to work on my essay, was the importance of dealing with the not-so-nice parts of the story of my place, the story of what has happened during my lifetime, as well as the story of the people who have lived here before me. When we think about our God as "the creator of heaven and earth," we so often think about God's creation as being perfect, "very good" indeed. But that which looks beautiful and bountiful has another side to it. The mountain we like to admire because of its beauty and magnitude can become the source of great atrocity and harm in a flash, for example, in the case of a volcanic eruption or an avalanche. What difference does that make to our understanding of God's creation? Is the mountain no longer a part of God's good and perfect creation when it has revealed its not-so-nice side to us?

3. **EMC and WJJ: In the title of this volume, there is a double wordplay. Firstly, there is not only an emphasis on the role (place) of story but also on the significance of place, the place where the story of God's work of creation is told. Such a place may provide a sense of being at home, but this place may itself be a "site of struggle." Could you please comment on such a sense of place?**

MLH: Appreciating the struggle of staying with the "problem of place," I am invited to consider the practices that emerge from our contemplative traditions that allow us to stay with the "struggle." What might it be like to have our hands in the dirt as we continue to think on the site of "Southern fruit trees" or lynching trees as part of the earth that deserve sacred devotion and even prayer?

EJ: I asked myself: Where is my place? My first place for my sense of being at home is my body. Our very bodies have been sites of patriarchal, racist, colonial, and ecological struggles. To maximize the concentration of wealth, workers have been set opposite each other as colonized, gendered, racialized, and deprived of natural means of subsistence. Women's bodies have been violated against their will. Black bodies have been kept in chains despite their revolt. Animals' bodies will be sold until humans evolve back to the millennia-old habit of consuming local fruits, vegetables, and whole grains. For those of us living under the neoliberal food regime, even our bodies are no longer natural but a chemical composition of less and less nutritious and more and more toxic foods, sourced by monocultural global farming. Anticipating our last question, our bodies are no longer natural or as initially created but rather capitalism's creation as fall.

LA: Any country may be our home. Even our bodies are not our homes. We rent the place, and our bodies are home to the Spirit of God. We are heavenly creatures, and our real home is in Heaven. Scripture says that Christians are citizens of Heaven, where the heavenly Father is King and Creator of all.

EMC: But Louk, does this not foster an escapist longing that was so vehemently criticized by Karl Marx and Friedrich Nietzsche? Should we not say, instead, that the earth is our God-given house but that it is not our home yet?

SC: My continual wrestling with the question of our place in the universe revolves around the issues of race, social class, and gender. This elicits concerns regarding the carbon footprint wrought by those who reside in the Global North and how this impacts heavily on residents of the Global South, whose carbon footprint is far lower. I am also painfully reminded of how those who are more economically endowed in the Global South are the ones engaging in activities which cause more ecological degradation to those who have not. Another concern at the forefront regarding our place in the universe has to do with gender justice, particularly how the earth is often perceived as female, "Mother Earth," and how the degradation suffered by Mother Earth is therefore twinly connected to the social, economic, and physical exploitation endured by women. Hence, these three "sites of struggle," race, social class, and gender, remind us of our ongoing striving to own our stories, to possess Mother Earth's bounty, and to love and be loved equally.

RA: I spoke mostly about treaty as a story that brings together the Creator, First People, and newcomers so that we could live like relatives in peace. All of this happens on the land, so land becomes, or remains, our relative who helps hold us together. There is struggle because there is distance between us. This distance we can respect and honor using ceremony or protocol. The space or the place, however, becomes the place of collaboration. A place where together we can do more than what we could do staying within our enclave. I see the incarnation as the goal, where creator and creation meet in perfect harmony in the person of Jesus Christ. What is fascinating is that Mary held creation and creator in perfect harmony within her womb. It seems to me that this could work to see how, within Mother Earth, she groans (Romans 8), longing to see us come to understand and live out this reality of harmony. I guess it just depends on how you tell the story...

AG: It is hard to see the stories of landlords and the landless, of those who have and those who have not, as one single story. Where does justice fit into the story of God's good creation? Would it be a story of God's creation if it wasn't about justice? Is the story of the creation of God a political story? Can it be something else?

WJJ: We need to stay with the problem of place. Place, as so many peoples understand it, is not simply their location, their place of habitation, but it is who they are. Place is the source of knowing and self-consciousness. Yet place, for so many of us, has been colonized by space. Space, in this regard, points to the history of mapping regimes of capital and property across places and silencing both place and people. The urgent question for us is: What does God's story say to this history?

4. EMC and WJJ: Pushing this further: there are different places at play in this volume, and one place may well be in conflict with other places near and far. How, then, does one tell a story of God creating the whole world from one specific place? How does one relate one such story with others?

SC: In response to this question, one of my favorite Scriptures, Psalm 24:1, "The earth is the Lord's and everything in it," comes to mind. It reminds us that the whole universe—whether it is the slums in Mbare, or the plush suburbs in Borrowdale, Harare, in Zimbabwe—both places were created by God at a certain time, for a specific purpose. The experiences of an artisanal miner on the outskirts of Zvishavane town or that of a professional miner at Mimosa Mine in Zvishavane, Zimbabwe, are also a constant reminder of our different places and different opportunities. Our places, stories, and experiences might be different; our fears, struggles, hopes and aspirations may be different, but that does not suggest that God, the Creator of the universe, views or values us any higher or lower. These diversities bring about a harmonious and melodious chord. Accepting our differences and respecting our different places and experiences will go a long way in helping us to live peacefully and harmoniously.

LA: Every place is different from one another, as every human being is different from one another. Even twin brothers have differences. Differences and diversities reveal the mighty and divine wisdom of God the Creator.

EMC: I am haunted by Vítor Westhelle's comment that where landlords see the beauty of God's creation, the landless only see gates and fences designed to keep them out.

EJ: Bodies of labor under imperialism employ the energy of their lives in projects that are not their own. Whereas a worker placed in the periphery of capitalism frequently sews a dress for the total cost of a quarter dollar, workers in centers of consumption resell the dress to earn many times more than that, and a consumer from an upper class finally pays for the dress, even as much as what the periphery worker will earn in a whole lifetime. There are places of freedom and places of slavery within the global chain of production. The majority are enslaved. Both the Arawete gods and the Christian God create life from death. Divine creation does not take place on Wall Street but in the midst of flesh expropriation, where life may be forged from death.

RA: I think I answered this earlier; for Indigenous people, we went through a time of violence. This is seen in the legends of the Haudenosaunee, and this is why the first ceremony is "meeting at woods' edge." You see the other at the woods' edge and go to them to offer them condolences. You comfort them, because how can you talk and come together if you do not first help people in their grief and pain? There is an understanding that we as humans have come through great difficulties and struggles. We are not trying to make everyone the same; we are trying to respect all people and work toward healing for the sake of our grandchildren. I think this is the desire for most, if not all, Indigenous people who have suffered, and this is where we find ourselves in this good world. The land or the place offered the opportunity to heal.

MLH: The gift of the volume, I think, is that it actually invites us, even places us in a position where we are confronted with the multiplicity of origin stories about creation. The essays invite us to de-center our own understanding of place and recognize that our theological imagination is being asked to stretch to incorporate more. Our sense of the one "holy space of God" is therefore shifting. There is a ground underneath that Paul Tillich and others have invited us to consider when engaging space and place.

5. **EMC and WJJ: Secondly, the title of this volume also suggests a tension between the place where a story is told and the story of that place—and contestations over how that story may be told. What do you make of that?**

MLH: Indeed, the aliveness of the tension is the birthplace of new ecotheologies.

LA: God created the world with a divine harmony and hierarchy, which is not to be interpreted as a hierarchy of power or possession. There should not be any tension or contestations if everything is interpreted in love and if humanity walks humbly before God. The problem is human greed for power, glory, and material possessions (Mic 6:8).

RA: I talk about Indigenous treaty because it flows from Indigenous law. Indigenous law, flowing from creation, teaches us we must seek to live in good relations with all people who come within our territory. Perhaps the local nature of place is lost when we use the word "creation," because we can hold so many stories and places in our heads!? Yet we live in a local place, and if we meet on the level of trying to help one another heal, perhaps that place can become a place of healing. We need creation and all within creation to help us because we can only see through the world we have known. My elders tell me, through their stories on the land, that it is a good world.

EJ: The patriarchal component in many world-known creation narratives seems to express the view that women are the property of men as much as humans are the property of the divine. Why have creation narratives often been told from the position of power of property holders? Is that why the most important gods in world religions are masculine figures?

SC: As an African ecofeminist theologian, I am always alert to the fact that my narrative as an African woman of faith and a gender activist academic is not always the same as those of my male counterparts, as well as those whose race and social class are different from mine. I am also attentive to my positionality as a middle-class African woman, which accords me a privileged position to tell my story, a privilege which might not be open for other women whose literacy is limited. Hence, in telling my story from my place, I do not claim to have walked the journey of those who have different experiences from mine. I, however, acknowledge that there are some common experiences which we might share, but I am also open to the fact that others might contest how I narrate my story.

WJJ: I appreciate Sophia's comments, because she situates herself in multiple struggles but also in multiple places to think of new possibilities. I seek to do something similar by thinking of the new places of identity and life from particular places that are yet contentious places.

AG: The emphasis on the contextualization of our theologies, so prevalent within the theological discourse for the past 50 years, has been invaluable, not the least because it has compelled us to recognize our limited point of view, as well as our biases. It has also helped us understand how important it is that we listen and learn from each other. Still, I worry that we tend to forget what we have in common, which I think is equally important. After all, the earth is "our common home" (as we are frequently reminded), which we are called to take care of. It is crucial that we find a balance between what we have in common, what is particular about our own context, and our own story. Our common experience of patriarchy is something all women share, but still, our experiences are very different, depending on our context. The same is true for the climate crisis. I don't think we will get anywhere if we don't admit our common responsibilities. At the same time, we need to face our different responsibilities, depending on where we live and how much we have contributed to the crisis. Not easy, but otherwise …

6. EMC and WJJ: How, then, would you respond directly to the core question of this volume: "*What difference does it make* to the story of cosmic, planetary, human, and cultural evolution to re-describe this as the creative work of God's love?"

EJ: Creation is a divine act. For the Arawete, humans do not even create songs. Only gods do. Plants used to exist already in paradise and were not created. When corporations claim rights of patent over seeds (Vandana Shiva), they are affirming themselves as gods. This is absurd for both Christian and Arawete faiths. For Christians, all life is God's creation. For the Arawete, all life, whether on Earth or in Heaven, has always relied on food from plants. Genetic variations are the result of thousands of years of cultivation and seed exchange among farmers and not the creation of any money-seeking human enterprise.

LA: The story of cosmic, planetary, human, and cultural evolution could be re-described as the creative work of God's love in the sense that God the Creator is offering an everlasting chance for humans to repent and to return to God the Creator. God gives to humans the freedom to learn from mistakes and the free gift of life in order to know God's divine love.

AG: The difference is between being its own maker (or creator) or being made (created) by something bigger or greater than itself. In other words, to look for a meaning in itself or to find that in a source outside of oneself.

SC: As we revisit the stories of our origins, we are reminded of our common humanity; we are also reminded of our interdependence as human and nonhuman entities. Our different stories resonate with the Genesis creation myths; they thread a beautiful tapestry, pointing us back to the Creator God, whose handiwork remains majestic and mysterious.

EMC: To answer the question directly, I think it makes a world of difference to describe this world as God's creation. In Cape Town, Abalimi Bezekhaya ("Planters of the Home") turns what is a virtual sand desert under electricity pylons into a lush oasis for small-scale urban farming. One may say that they see such degraded land as nothing but God's own vegetable garden.

RA: Our creation story tells us how we are related to the earth. We are the most dependent creatures, but we have been given the responsibility to draw forth what is best from the earth. Mission and ministry are about joining with the Creator's work to bring all things into harmony. I thought if we joined with the creative work of the Creator, this would help revive a sense of the common good.

7. EMC and WJJ: This volume seeks to optimize diversity in ecotheology. Amid such diversity, what "current paths" are you able to identity in the field? And what "emerging horizons"? Where is the debate going?

EJ: A concrete path that I see for religious communities is gardening. All diversity of ideas is welcome while gardening and eating together.

LA: A concrete path is the recognition of Indigenous spirituality and opposition to the Western trends of understanding the value of life. Instead, we may value nature and relationships within a framework of interdependency. Abstinence to monetary value and economic growth as an indicator of well-being is then required.

SC: The menacing ecological crisis reminds us of our frailty and our finite nature. In my view, the important issues to be addressed by ecotheologians include paying attention to indigenous ways of preserving Mother Earth, acknowledging the inseparable connection between the ecological devastation of Mother Earth and women, redressing the impact of the ecological devastation caused by the Global North on the Global South, and acknowledging that there can never be any lasting peace and economic progress in the midst of an ecological crisis.

AG: It is hard to imagine an ecotheology which does not address the current climate crisis. Even if the consequences of a warming climate are not as visible everywhere, global warming poses a threat to the future of all life on this planet. This is why melting ice in the Arctic is not only threatening to life in the northernmost part of Earth, but also in the Southern Hemisphere. As the saying goes: "What happens in the Arctic doesn't stay in the Arctic."

This is an important reminder of the interconnectedness of all parts within God's creation, hence the interdependence between all our stories.

EMC: Perhaps the obvious needs to be stated given the shifts (or "current paths") in approaching creation theology outlined in the editorial introduction: God's work of creation cannot be reduced to theories of origination or to "the beginning," although it also does not exclude that. The sense of place that is so overwhelmingly evident in the essays in this volume suggests that describing the world as God's creation is about the local and the present. Only on that basis can "horizons" be extended to the "ends of the Earth," to use a biblical phrase.

RA: My chief told me, "the only thing a person can possess is what they can carry on their back. So how could you own the land?" The land carries us, so we ought to respect the earth, like you would respect your mother. The great opportunity we have is to see that we have to work together to heal the land. We already do that when it comes to floods and forest fires. Everyone works together because we can see that something needs to be done. The challenge is to continue to create a conversation about the current climate challenges. The gospel reminds us that this is what we have been sent to do.

8. EMC and WJJ: If you were to give advice to people who are entering into ecotheology or to people who are seeking to root God's work of creation in a particular place, what would that advice be?

EJ: Start a community garden! To root divine creation in a place means to have that place become independent from international corporate control in food production.

SC: Cherishing Mother Earth and doing our part to reclothe and continually restore Mother Earth to its original state should be our goal. Responsible consumption of natural resources, recycling, reusing, and restoring nature's bounty must be good news for God's creation. Eating our food as medicine, exercising, and maintaining a healthy and responsible work ethic should be our clarion call.

LA: Stop breathing and close your eyes for as long as you can. You will faint if you continue, or you will have to open your nose, mouth, and eyes to continue to live. Life is a divine gift of God the Creator, and every part of creation (human, air, light, etc.) has to be in communication with each other in an ecological relationship of interdependence.

RA: I would tell them that you need to learn how to listen to everyone so that you can put into a narrative a consensus in which everyone feels they were heard. Even if they do not agree completely with what is happening going forward, they feel they can join in because it is a story that holds everyone together. We need to create a social contract for the common good around creation or the land. A late friend of mine said that we have mostly believed the right things, but we have believed too small. Tell the story in such a way that everyone fits in, and you will change the world.

9. EMC and WJJ: Given what we know about the so-called Anthropocene, do you think it is appropriate to describe this world as God's creation? Or is the world now "our" creation? How should the rest of the story then be told—preempting the subsequent volumes in this series?

RA: This last question reminds me that in our modern world, we continue to see ourselves as separate from creation. We are estranged from creation, but the journey to understand that the earth is our relative opens up many possibilities. To see again that Christ taking a human body in which creation and the Creator exist in perfect harmony points us in the right direction. We will impact creation and creation will impact us; it has always been thus. We must learn to live in harmony. This is the gospel, and this is the task of becoming a true human being.

EJ: Marx and Engels write that nature no longer exists, unless in a few recent coral islands (*German Ideology*). Nature continues to create, the earth continues to bring forth grass (Gen 1:11), and the waters still give life to moving creatures (Gen 1:20), but that self-creation now finds itself under deadly capitalist control. Capitalism exploits nature for profit and not with the purpose of maintaining life. Rivers are contaminated, seeds are genetically modified, supermarket foods are filled with sugar and palm oil, and animals are left without habitat due to deforestation.

LA: The "Anthropocene" means that human activities affect the main characteristics of the earth. It is not appropriate to say that the world is now "our" human creation, because God the Creator is still holding the future and the basis of life. Humans cannot make "seeds" or "eggs" from anything. Humans can only modify or alter the existing creation by attempting to manipulate genes or the divine natural laws for creation. The rest of the story depends on the reconciliation of humans with the divine Creator, and that is one of the goals of this volume.

EMC: In my essay, I argued that the confession that the world is God's creation is not to be taken for granted but is indeed deeply counterintuitive and polemic. This challenges the "Anthropocene" claim that the earth is in human hands (David Grinspoon). How this story is to be told remains open-ended, so the jury is still out!

AG: Because it is so central to our Christian faith to confess that the world is God's creation, the danger is that we do not pay much attention to what it means or could mean today. When key ideas of our faith, like the one about the world as God's creation, have become cliché, they have lost their ability to be counterintuitive and polemic.

SC: The earth remains God's gift to humanity. We are just sojourners on Mother Earth. The best we can do right now is to try our level best to reduce the collective carbon footprint. What we give is what we get; it is never too late for us to stop pollution, riverbank cultivation, littering, desertification, wildfires, poaching, and all that which destroys and diminishes Mother Earth.

WJJ: The question captures the profound risk that God has entered with us, God's own creation. That risk is co-creating through life bound together. This is what the incarnate God shows us. A God willing to enter in the making with us, seeking through the making to draw us toward the divine life and a thriving life together. This, in fact, is what we mean by grace.

AG: Arnfríður Guðmundsdóttir

EJ: Eneida Jacobsen

EMC: Ernst M. Conradie

LA: Louk Andrianos

MLH: Melanie L. Harris

RA: Ray Aldred

SC: Sophia Chirongoma

WJJ: Willie James Jennings

Index

A
Aldred, Raymond C., 25, 27, 29, 31, 33, 35, 37, 39, 171, 183
apartheid theology, 5, 13, 84-86, 89
Arawete theology, 139-140, 142, 150-152
avalanches, 102, 105, 114-118

B
Bavinck, Herman, 5, 84, 87, 94, 98-100
Berry, Thomas, 6, 9, 24, 83, 89
Billmann, Kathleen, 116, 118
Black theology, 15-16, 23, 125
blue cross movement, 42, 47-49

C
cannibalism, 143, 151-152
capitalism, 5, 14, 137, 139, 148-153, 161, 175, 177, 182
Chirongoma, Sophia, 17, 59-61, 63, 65, 67-71, 73-78, 108, 110, 114, 118, 171, 179, 183
compassion, 116-118, 132, 134
Cone, James, 1, 7, 9, 14, 22-24, 90, 98-99, 112, 118, 124-126, 128, 136, 146, 153, 155, 168-169, 174, 183
Conradie, Ernst M., 1-3, 5-11, 13-17, 19, 21-23, 25, 41, 59, 75, 77, 79, 81, 83, 85-89, 91-93, 95-101, 121, 137, 155, 171, 183

D
Dzivaguru creation myth, 60-61, 63-65

E
ecofeminist theology, 12, 24, 61, 72, 154
ecumenical theology, 22, 99

F
Farley, Wendy, 116, 118
feminist theology, 77
Fihavanana, 45-47, 57-58
Flateyri, 114-115, 118
food production, 73, 138, 160, 181

G
Gebara, Ivone, 13, 15, 23
Guðmundsdóttir, Arnfríður, 72, 101, 103, 105, 107, 109, 111-113, 115-119, 171, 183
Guruuswa creation myth, 60, 65-66

H
Harris, Melanie, 15, 23, 72, 121, 123, 125, 127, 129, 131, 133, 135-137, 166, 168, 171, 183

I
Iceland, 72, 101-104, 106-109, 113-115, 117-119
indigenous theology, 25-26, 46

J
Jennings, Willie James, 1, 3, 5, 7, 9, 11-13, 15-17, 19, 21, 23, 25, 41, 59, 79, 86, 89, 99, 101, 112, 118, 121, 131, 136-137, 155-159, 161-163, 165-169, 171, 183
Johnson, Elizabeth, 7, 11-12, 23, 76, 111, 115-116, 118, 124-126, 128, 136, 174

K
Kuyper, Abraham, 79, 84-87, 99

L
laments, 111-112
liberation theology, 13, 15, 122, 138
Luther, Martin, 4, 28, 39, 110, 112, 116-118

M
McFague, Sallie, 7, 11-12, 23, 83, 111, 118
Migliore, Daniel, 13, 22, 116, 118
Moltmann, Jürgen, 4, 6-7, 9, 23, 84, 87, 94, 96, 100
Mwedzi creation myth, 62-63

N
natural disasters, 68-69, 74, 76, 101-102, 104-106, 108, 110, 112-118, 174
Neo-Calvinism, 84, 87, 94

O
Orthodox Academy of Crete, 42, 49-50, 52-53

P
patriarchy, 10-11, 63, 65, 76-77, 84, 89, 130, 137, 145, 148, 150-153, 161, 169, 179

S
Shona people, 59-62, 64-66, 68, 70, 72, 74-76, 78
Southgate, Christopher, 9-10, 22, 24, 129
Steingrímsson, Jón, 106-111, 114, 117-119
Súðavík, 114-115, 118

T
Thompson, Deanna, 32, 112, 118-119

Index

V
Van Ruler, Arnold, 5, 84, 87-88, 100
Viveiros de Castro, Eduardo, 137, 150, 153-154, 169
volcanic eruptions, 72, 101-102, 104-106

W
Western theology, 25-26
Westhelle, Vitor, 13, 24, 89, 100, 177

womanist theology, 122-123
World Council of Churches, 8, 17, 41-42, 53, 91

Z
Zimbabwe, 59-61, 63, 68-69, 71, 74-78, 108, 118, 176-177